普通高等学校计算机类专业特色教材·精选系列

Java EE 项目开发与设计实验指导书

彭灿华　韦晓敏　杨呈永　主编

中国铁道出版社
CHINA RAILWAY PUBLISHING HOUSE

内 容 简 介

本书是《J2EE项目开发与设计（第二版）》的配套实验指导书，其中所有实验均与主教材《J2EE项目开发与设计（第二版）》配合进行，包括使用MySQL作为后台数据库进行J2EE（已更名为Java EE）项目开发的方法。

本书适合作为高等院校计算机相关专业的实验教材或教学参考书，也可作为相关培训机构的教材及软件设计人员的辅导用书。

图书在版编目（CIP）数据

Java EE项目开发与设计实验指导书/彭灿华，韦晓敏，杨呈永主编.—北京：中国铁道出版社，2017.9
普通高等学校计算机类专业特色教材．精选系列
ISBN 978-7-113-23445-4

Ⅰ.①J… Ⅱ.①彭… ②韦… ③杨… Ⅲ.①JAVA语言－程序设计－高等学校－教材 Ⅳ.①TP312.8

中国版本图书馆CIP数据核字（2017）第181849号

书　　名：Java EE项目开发与设计实验指导书
作　　者：彭灿华　韦晓敏　杨呈永　主编

策　　划：祝和谊　　**读者热线**：（010）63550836
责任编辑：周　欣　卢　笛
封面设计：一克米工作室
封面制作：刘　颖
责任校对：张玉华
责任印制：郭向伟

出版发行：中国铁道出版社（100054，北京市西城区右安门西街8号）
网　　址：http://www.tdpress.com/51eds/
印　　刷：虎彩印艺股份有限公司
版　　次：2017年9月第1版　2017年9月第1次印刷
开　　本：787 mm×1 092 mm　1/16　**印张**：4.5　**字数**：101千
书　　号：ISBN 978-7-113-23445-4
定　　价：15.00元

前　言

本书内容

本书是《J2EE 项目开发与设计（第二版）》的配套实验指导书，其中所有实验均与主教材《J2EE 项目开发与设计（第二版）》同步进行，包括使用 MySQL 作为后台数据库进行 J2EE（已更名为 Java EE）项目开发的方法。

主教材《J2EE 项目开发与设计（第二版）》是笔者在多年项目开发过程中的经验总结，通过丰富的实例，由浅入深、循序渐进地介绍了目前采用 Java 进行 Web 开发的各种框架的使用方法，从而帮助软件设计人员快速掌握 Web 开发技术，并能将其应用在实战中。

本书由实验效果预览、工程目录结构说明和十一个具体实验组成。

本书由彭灿华、韦晓敏、杨呈永担任主编。

本书特色

本书详细列出每个案例的开发步骤，实例易于阅读和理解。综合案例以软件工程的标准进行设计并开发，编程理念面向需求、面向市场。

本书适用对象

本书适合作为高等院校计算机相关专业的教材或教学参考书，也可作为相关培训机构的教材及软件设计人员的辅导用书。

由于水平有限，书中疏漏之处在所难免，恳请读者批评指正。读者有任何意见与建议或者在学习的过程中遇到不解的地方，都可以通过邮件与编者进行探讨，也可通过邮件索取本书源代码及相关视频。

联系方法如下：

电子邮箱：449271349@qq.com。

编者

2017 年 5 月

目 录

实验效果预览

通过本书中的实验，将实现一个完整的问卷在线提交系统，系统预览效果如下。

前台效果预览：

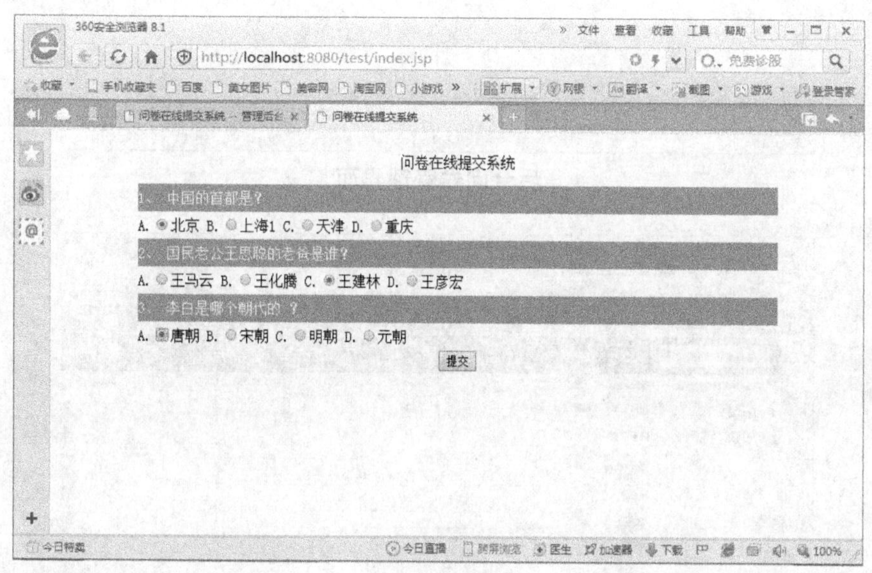

前台界面

后台效果预览：

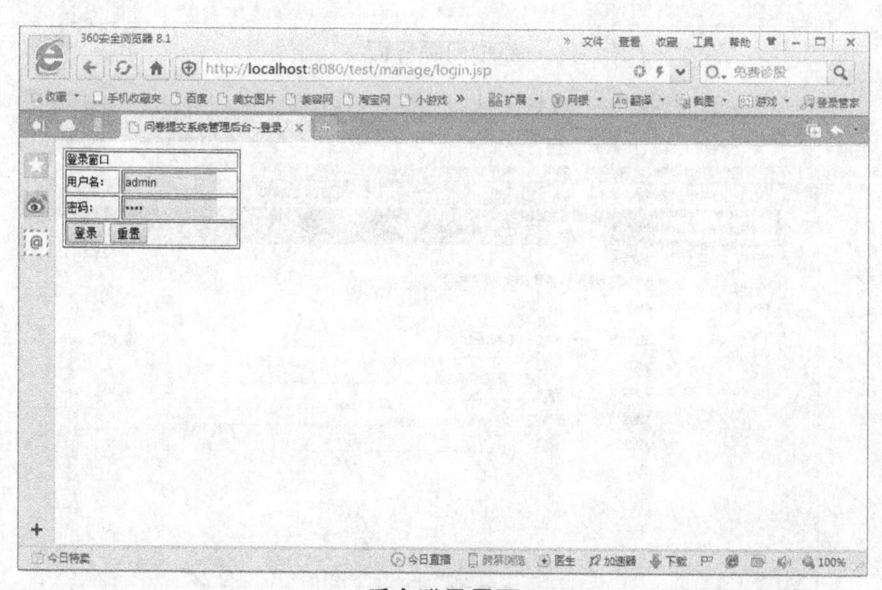

后台登录界面

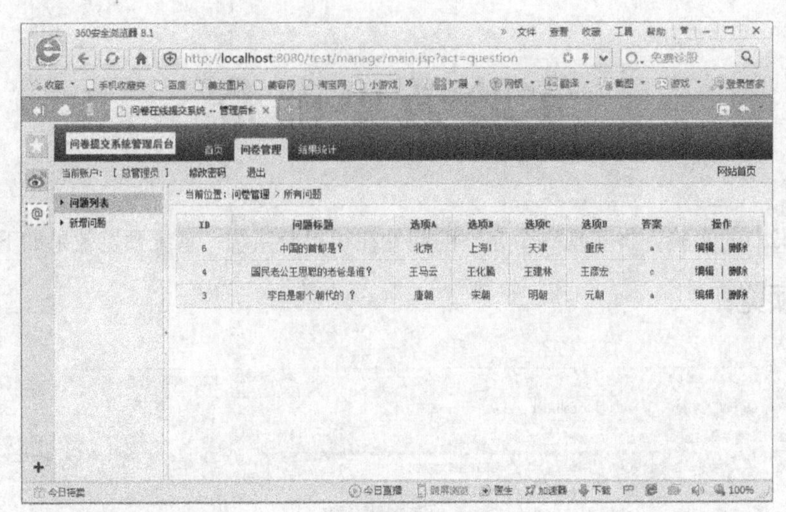

后台问卷管理界面

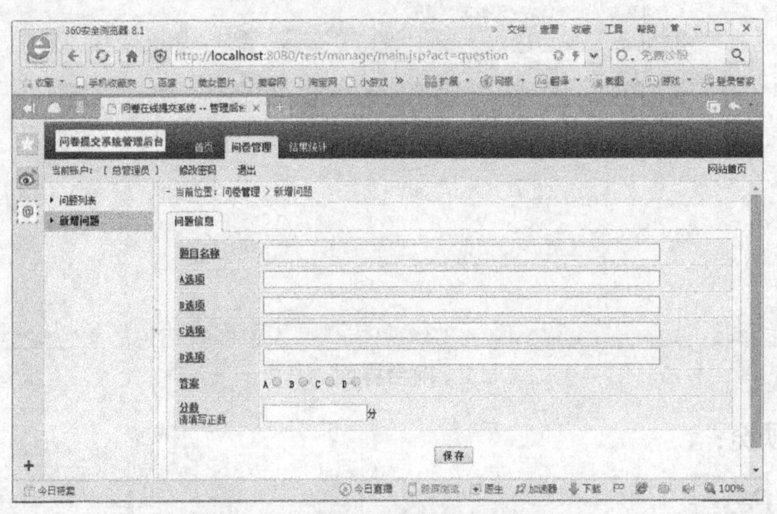

新增问题界面

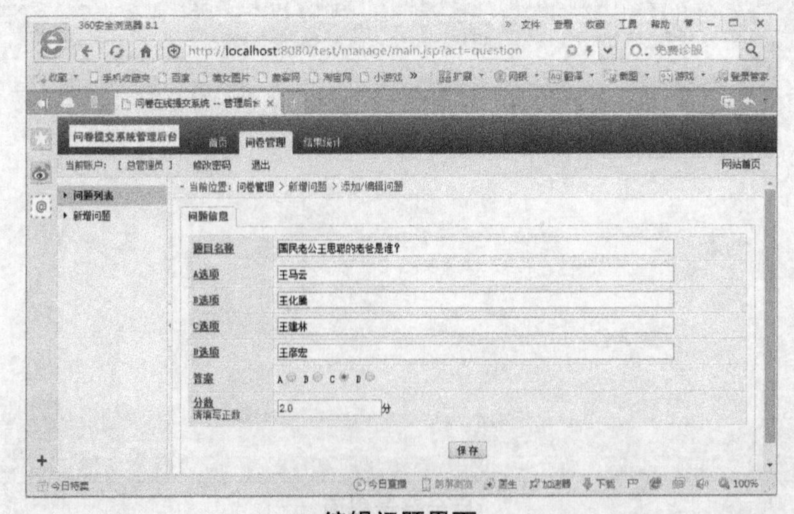

编辑问题界面

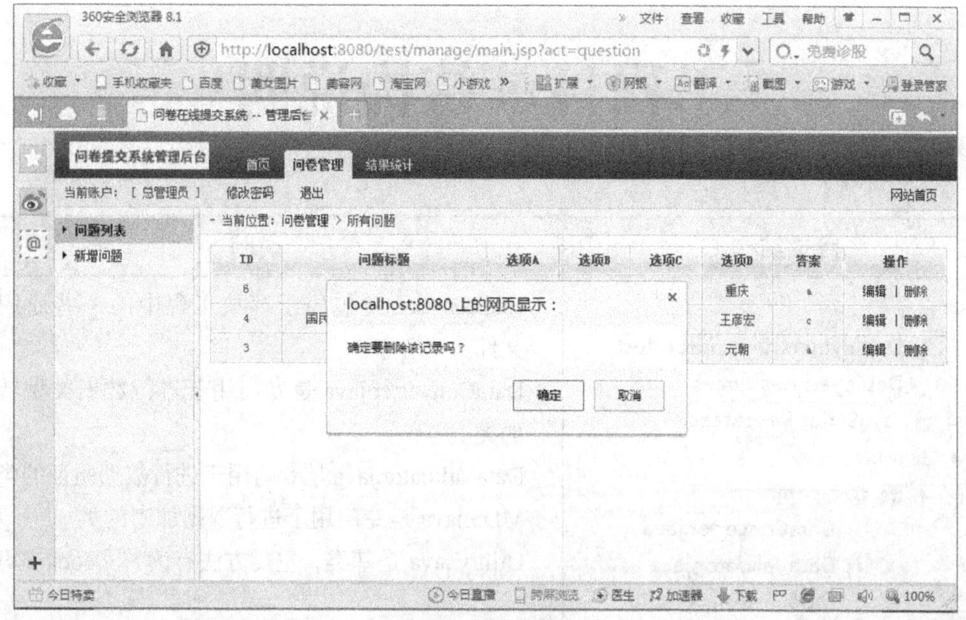

删除问题界面

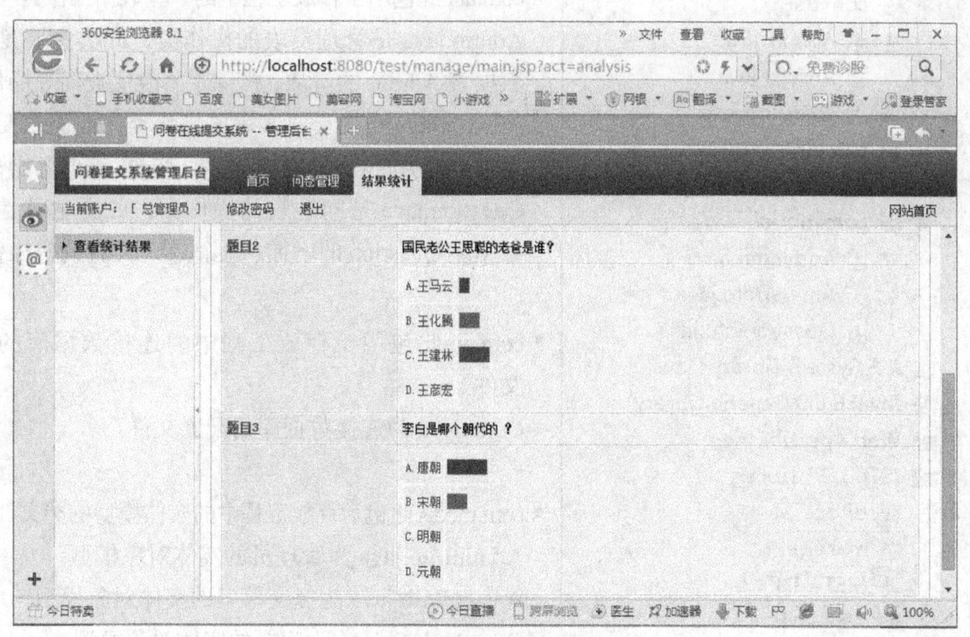

提交结果统计界面

工程目录结构说明

目录结构	说明
test Deployment Descriptor: test Deployed Resources JavaScript Resources src com.common DataConverter.java DataValidator.java MD5.java Utility.java	• com.common 此包用于存放工程中的一些通用类文件。 DataConverter.java 是专门用于进行数据类型转换的类。 DataValidator.java是专门用于进行数据验证的类。 MD5.java 是专门用于进行数据加密的类。 Utility.java 是基类，主要方法有读 / 写 cookie 等。
com.dal Admin.java Answer.java Question.java	• com.dal 此包用于存放工程中的一些数据操作类。 Admin.java 是管理员表的操作类，此类主要是对数据表 Admin 的增加、删除、修改、查询等操作。 Answer.java 是提交答案表的操作类，此类主要是对数据表 Answer 的增加、删除、修改、查询等操作。 Question.java 是问题表的操作类，此类主要是对数据表 Question 的增加、删除、修改、查询等操作。
com.db Conn.java	• com.db 此包用于存放工程中的连接数据库的类文件。 Conn.java 为连接数据库的配置文件。
com.model AdminInfo.java AnswerInfo.java QuestionInfo.java JRE System Library [JavaSE-1.6] JavaEE 6.0 Generic Library Web App Libraries JSTL 1.2.1 Library	• com.model 此包为存放工程中的实体模型的类文件。 AdminInfo.java 为管理员的实体对象模型。 AnswerInfo.java 为提交答案的实体对象模型。 QuestionInfo.java 为问题的实体对象模型。
WebRoot manage META-INF WEB-INF index.jsp result.jsp save.jsp	• WebRoot/manage 为管理后台根目录。

实验一 JSP 开发环境的配置

实验目的

☑ 掌握 Java EE 开发环境的搭建。
☑ 掌握 MyEclipse 的安装与使用。
☑ 掌握 MyEclipse 开发工具的使用与优化。

实验要求

（1）将本次实验的所有工程文件保留。

（2）实验结束后，按要求完成实验报告。

实验课时

2 课时。

实验准备

（1）准备所需软件（在实验课前会发放给大家），如表 1-1 所示。

表 1-1　所需软件

软件名称	版本号	说明	下载地址
MyEclipse	2015	myeclipse-2015-offline-installer-windows	https://downloads.myeclipseide.com/downloads/products/eworkbench/2015/installers/myeclipse-2015-2014-07-11-offline-installer-windows.exe
操作系统	Windows 7/10	32 位 / 64 位	

（2）复习 HTML 网页元素的组成。

实验内容

（1）安装 MyEclipse，如图 1-1 所示。为方便后面的使用，建议将 MyEclipse 安装在 C 盘根目录下。

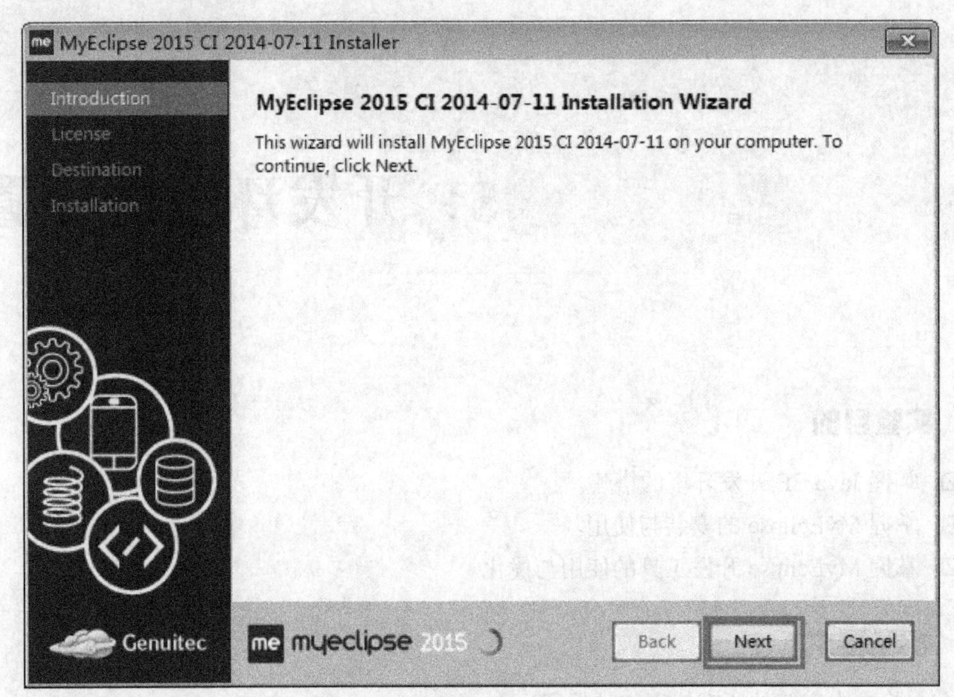

图 1-1　MyEclipse 2015 安装界面

（2）新建第一个 Web Project，如图 1-2 ～图 1-4 所示。

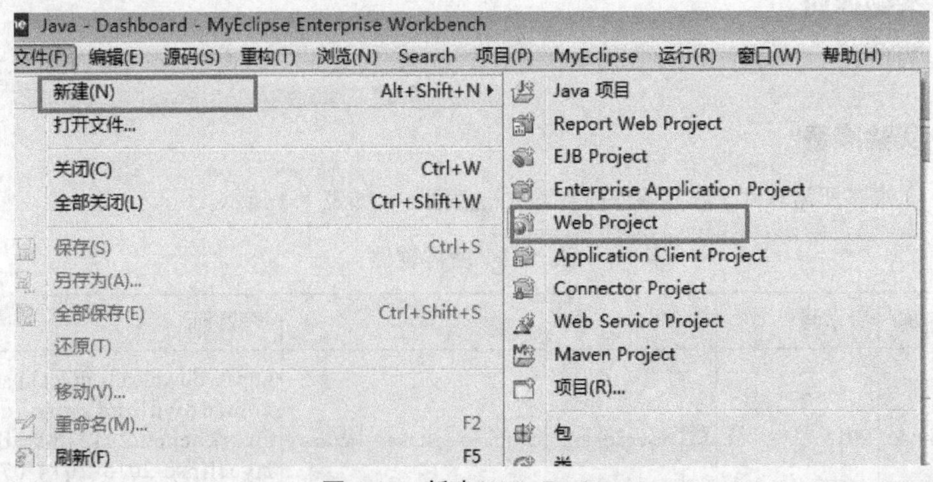

图 1-2　新建 Web Project

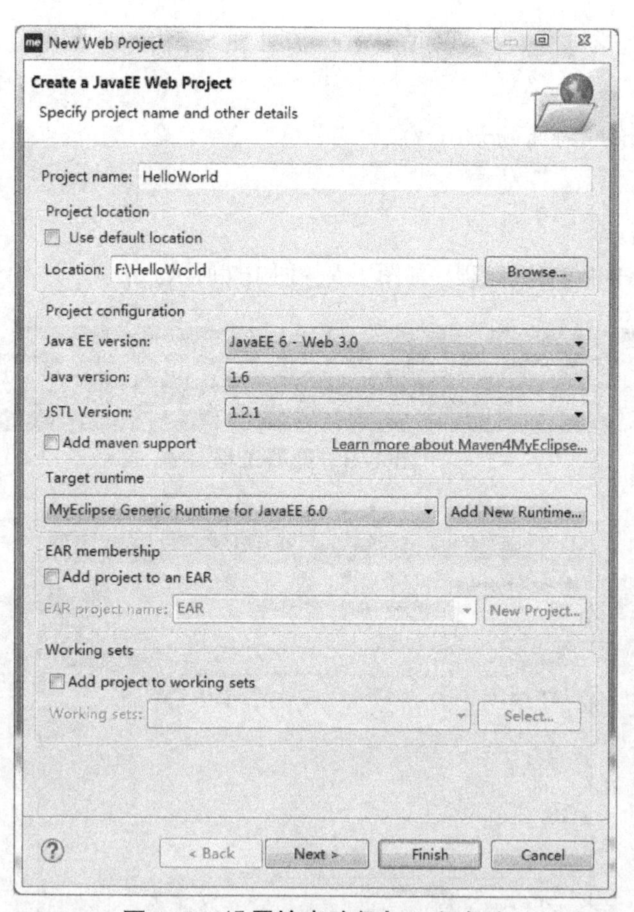

图 1-3　设置输出路径与工程名称

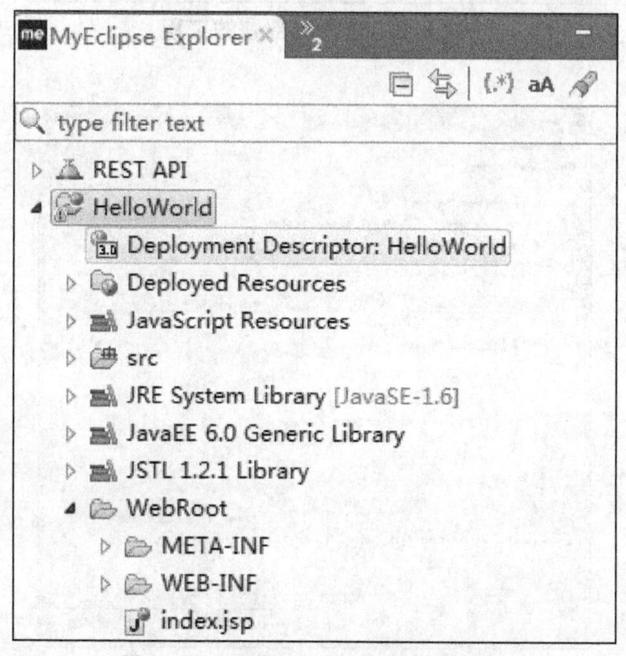

图 1-4　HelloWorld 工程目录结构

双击 index.jsp，找到 <body></body> 标签，在此标签中输入如下代码。

```
<%
 out.println("Hello World!");
%>
```

（3）部署运行。

第一步：部署 HelloWorld 工程，如图 1-5 ～图 1-7 所示。

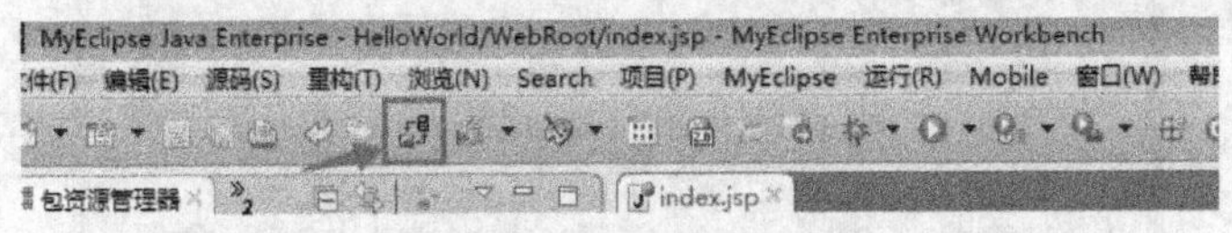

图 1–5　部署工程

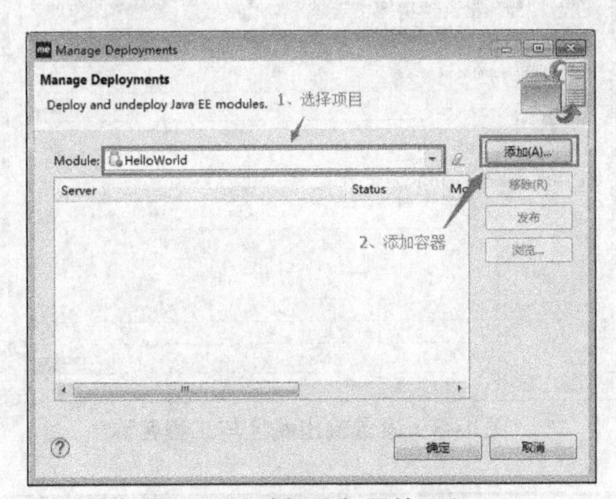

图 1–6　选择要部署的工程

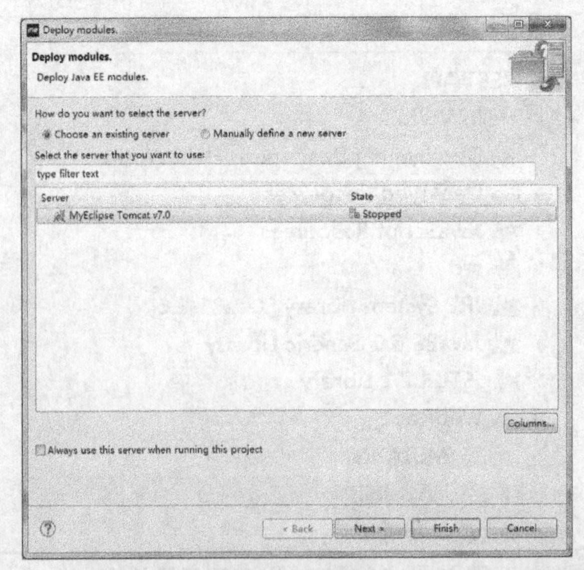

图 1–7　选择部署服务器使用 MyEclipse 2015 自带的服务

第二步：启动 Tomcat v7.0，如图 1-8 所示。

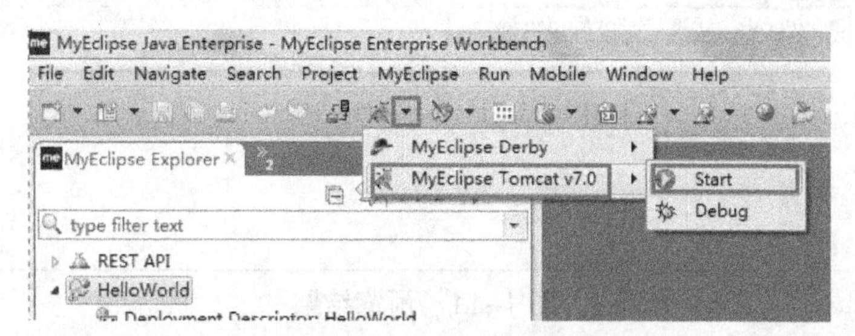

图 1-8　启动 Tomcat v7.0

服务启动后，查看启动日志，检查是否有异常，如图 1-9 所示。

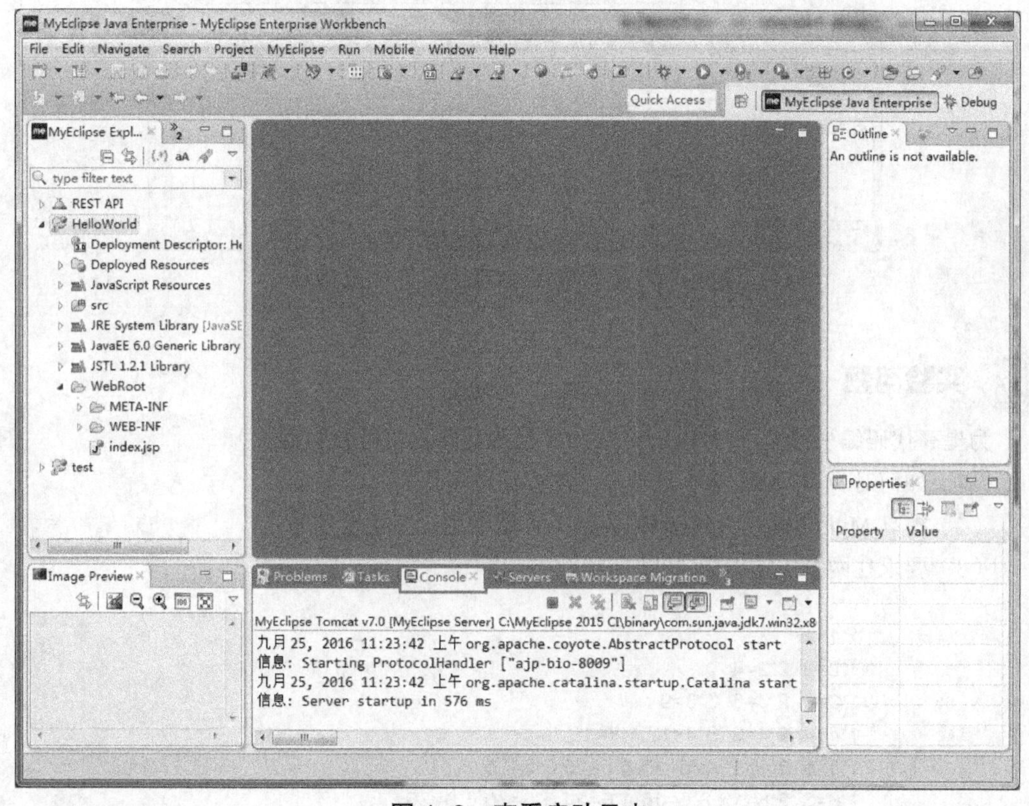

图 1-9　查看启动日志

第三步：预览结果，如图 1-10 和图 1-11 所示。

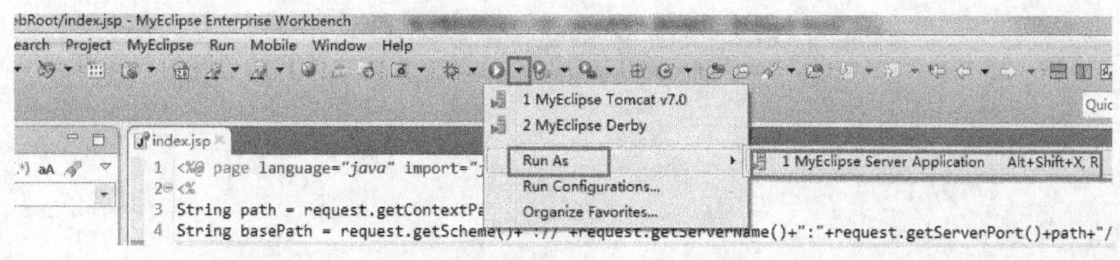

图 1-10　预览 index.jsp 页面

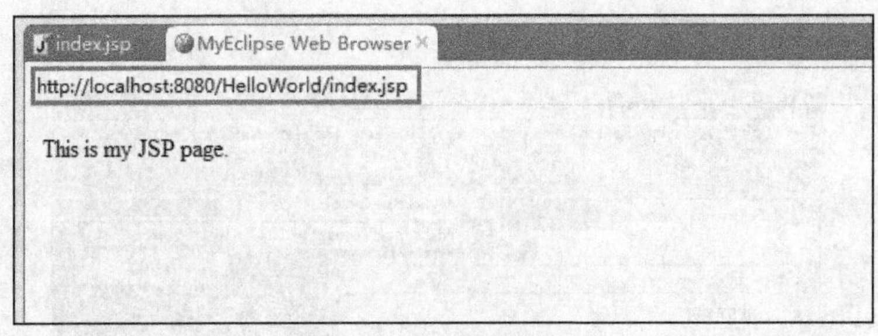

图 1-11 预览结果

也可复制图 1-11 中的 URL 地址，在浏览器中直接打开，如图 1-12 所示。

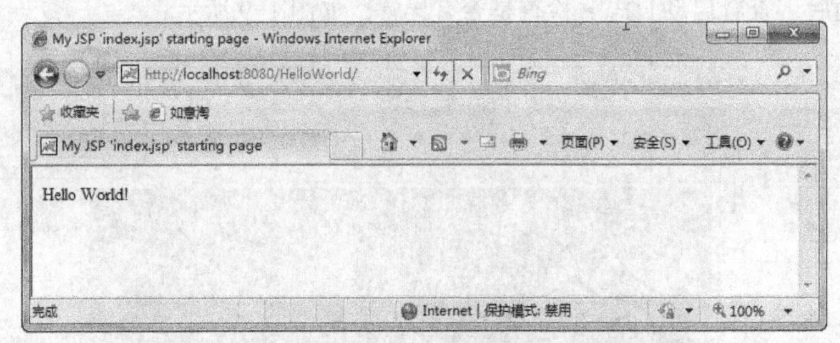

图 1-12 在浏览器中直接打开指定地址

实验习题

为提高代码编写速度，MyEclipse 有内容提醒功能。（可参照附录）

（1）如何对 MyEclipse 进行注册。（可参照附录）

（2）提高 MyEclipse 每行代码的容量，方便使用。（可参照附录）

（3）要求在网页中输出图 1-13 所示效果。

```
1*1=1
1*2=2  2*2=4
1*3=3  2*3=6  3*3=9
1*4=4  2*4=8  3*4=12 4*4=16
1*5=5  2*5=10 3*5=15 4*5=20 5*5=25
1*6=6  2*6=12 3*6=18 4*6=24 5*6=30 6*6=36
1*7=7  2*7=14 3*7=21 4*7=28 5*7=35 6*7=42 7*7=49
1*8=8  2*8=16 3*8=24 4*8=32 5*8=40 6*8=48 7*8=56 8*8=64
1*9=9  2*9=18 3*9=27 4*9=36 5*9=45 6*9=54 7*9=63 8*9=72 9*9=81
```

图 1-13 “九九乘法表”效果显示

实验二
JSP对象的使用——设计简易问卷提交系统

实验目的

☑ 掌握 JSP 的基本语法：JSP 注释、JSP 脚本元素、JSP 指令、JSP 动作指令。
☑ 掌握 JSP 内建对象。

实验要求

（1）本实验要求在实验一的基础上进行。
（2）将本次实验的所有工程文件保留。
（3）实验结束后，按要求完成实验报告。

实验课时

2 课时。

实验准备

（1）本实验需新增表 2-1 所示的软件（在实验课前会发放给大家）。

表 2–1　新增软件

软件名称	版本号	说明
Dreamweaver	CS6	Dw Adobe Dreamweaver CS6

（2）复习 HTML 中表单元素的使用。

实验内容

（1）新建一个 Web Project 工程，在此工程中新建一个 jsp 页面，命名为 Question.jsp，核心代码如下所示：

```
<FORM action="Answer.jsp" method="post" name="form1">
诗人李白是中国历史上哪个朝代的人：<BR>
        <INPUT type="radio" name="R" value="a"> 宋朝
        <INPUT type="radio" name="R" value="b">唐朝
        <INPUT type="radio" name="R" value="c"> 明朝
        <INPUT type="radio" name="R" value="dv > 元朝
         <BR>
<P> 小说红楼梦的作者是：
         <BR>
        <INPUT type="radio" name="P" value="a">曹雪芹
        <INPUT type="radio" name="P" value="b"> 罗贯中
        <INPUT type="radio" name="P" value="c"> 李白
        <INPUT type="radio" name="P" value="d"> 司马迁
        <BR>
        <INPUT TYPE="submit" value=" 提交答案 " name="submit">
    </FORM>
```

（2）在此工程中新建一个 jsp 页面，命名为 Answer.jsp，此页面要求显示图 2-1 所示效果。

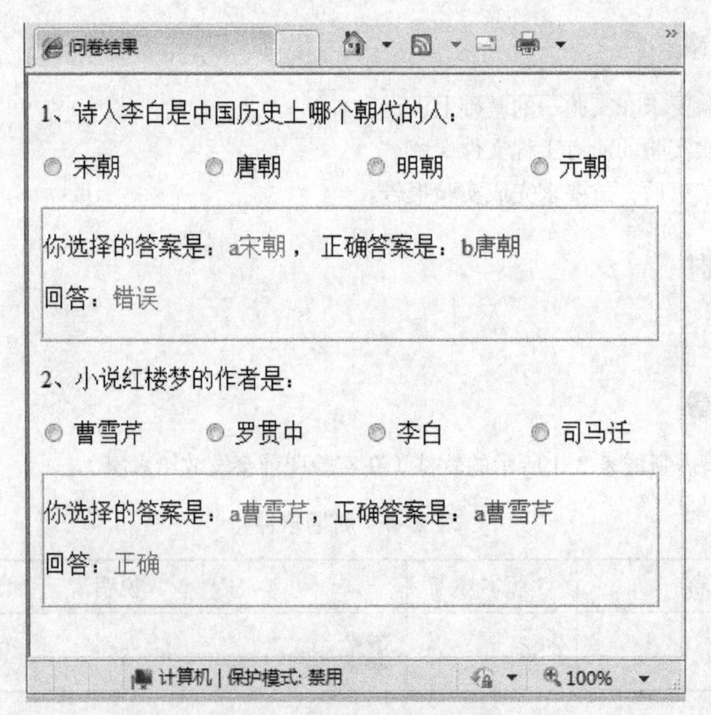

图 2-1　Answer.jsp 页面

实验三

JSP 对象的使用——实现问卷提交系统管理后台界面

实验目的

☑ 掌握 JSP 的基本语法：JSP 注释、JSP 脚本元素、JSP 指令、JSP 动作指令。

☑ 掌握 JSP 内建对象。

实验要求

（1）将本次实验的所有工程文件保留。

（2）实验结束后，按要求完成实验报告。

实验课时

2 课时。

实验准备

复习 HTML 中 iframe 框架的使用。

实验内容

（1）新建一个 Web Project 工程，在此工程中新建一个目录，命名为 manage。在此目录下新建一个 jsp 页面，命名为 login.jsp，预览效果如图 3-1 所示。

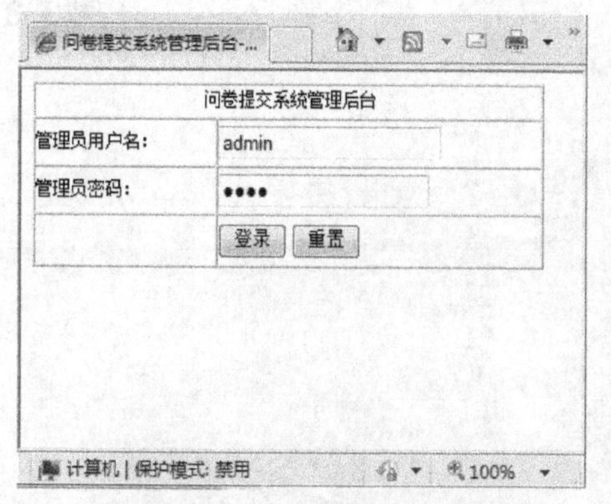

图 3–1　login.jsp 预览效果

（2）在此 Web Project 工程中，新建一个 jsp 页面，命名为 main.jsp，包括顶部的 top.jsp，左边的是菜单栏 left.jsp，右边的是主页 right.jsp，目录结构如图 3-2 所示。

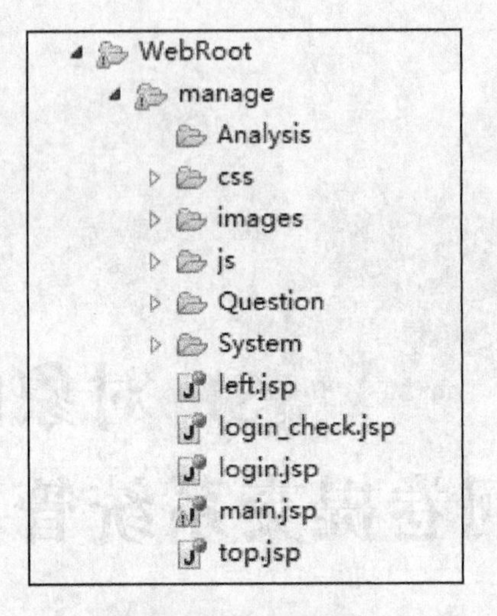

图 3-2 目录结构

main.jsp 的预览效果如图 3-3 所示。

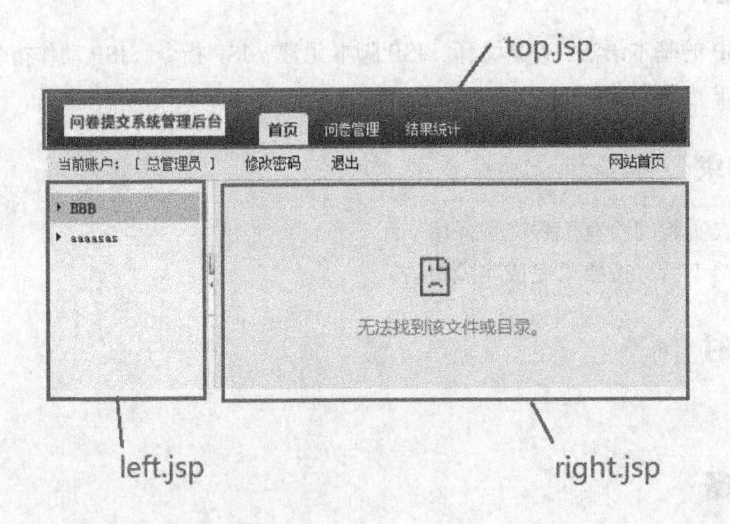

图 3-3 main.jsp 预览效果

实验四

JavaBean 的使用——后台登录模块使用 JSP+JavaBean

实验目的

☑ 了解什么是 JavaBean。
☑ 掌握 JavaBean 的开发与使用。
☑ 掌握 JavaDoc 文档的生成。
☑ 掌握 Jar 插件的制作与使用。

实验要求

（1）本实验要求在实验三的基础上进行。
（2）将本次实验的所有工程文件保留。
（3）实验结束后，按要求完成实验报告。

实验课时

2 课时。

实验准备

复习 session 的含义及其作用。

实验内容

（1）打开 login.jsp 页面（见图 3-1），将表单元素进行重命名。核心代码如下：

```
<form id="form1" name="form1" method ="post" action="<%=basePath%>/manage/login_check.jsp?action=LoginAction">
        <table width="199" border="1">
```

```
                <tr>
                    <td colspan="2">登录窗口</td>
                </tr>
                <tr>
                    <td>用户名:</td>
                    <td><input name="txtUsername" type="text" size="12" /></td>
                </tr>
                <tr>
                    <td>密码:</td>
                    <td><input name="txtPassword" type="password" size="12" /></td>
                </tr>
                <tr>
                    <td colspan="2"><input type="submit" name="button" id="button"
value="登录" /> <input type="reset" name="button2" id="button2" value="重置" /></td>
                </tr>
            </table>
        </form>
```

（2）新建 login_check.jsp 页面，实现用户名与密码的登录验证。假设正确的用户名为“admin”，密码为“1234”，当输入此组信息后，进入 main.jsp 页面。否则显示“用户名密码错误”。核心代码如下：

```
<%@ page language="java" import="java.util.*" pageEncoding="utf-8"%>
<%
if ("LoginAction".equals(request.getParameter("action")))
{
    String txtUsername = request.getParameter("txtUsername");
    String txtPassword = request.getParameter("txtPassword");
    //假设正确的用户名与密码是 admin ,1234
    //接下来判断用户输入的用户名与密码是否正确
    if ("admin".equals(txtUsername) && "1234".equals(txtPassword))
    {
        session.setAttribute("CurrentUser",txtUsername);
        response.sendRedirect("main.jsp");
    }
    else
    {
        out.print("<script>alert('用户名密码不正确');window.loaction.href
('login.jsp');</script>");
    }
}
else
```

```
    {
        out.print("<script>alert('非法访问!');window.loaction.href('login.
jsp');</script>");

    }
    %>
```

实验练习

（1）将 login_check.jsp 文件内容改为 JavaBean 文件。

（2）将 JavaBean 文件制作成 jar 文件，在此工程中使用。

实验五
JSP+MySQL 查询操作实现后台登录验证从数据库判断

实验目的

☑ 掌握 MySQL 数据库及 Navicat for MySQL 的使用。

☑ 熟练使用 SQL 语句进行数据查询、数据表创建、数据库创建。

☑ 掌握 JSP 连接 MySQL 数据库的方法与步骤。

☑ 掌握在浏览的开发模式下查看 cookie 的方法。

实验要求

（1）本实验要求在实验四的基础上进行。

（2）将本次实验的所有工程文件保留。

（3）实验结束后，按要求完成实验报告。

实验课时

2 课时。

实验准备

本实验新增表 5-1 所示软件（在实验课前会发放给大家）。

表 5-1　新增软件

软件名称	版本号	说明
MySQL 连接驱动	5.1.6	mysql-connector-java-5.1.6-bin.jar

（1）复习常用的数据库操作语句。

（2）复习 cookie 的含义及使用方法。

（3）复习 Java 中对象的使用和类方法的调用。

实验内容

（1）安装 MySQL，安装步骤参照主教材（特别注意：设置数据库用户名与密码）。

（2）安装 Navicat for MySQL 可视化操作软件。

（3）在 Navicat 中新建一个数据库，命名为 QuestionDB，如图 5-1 所示。

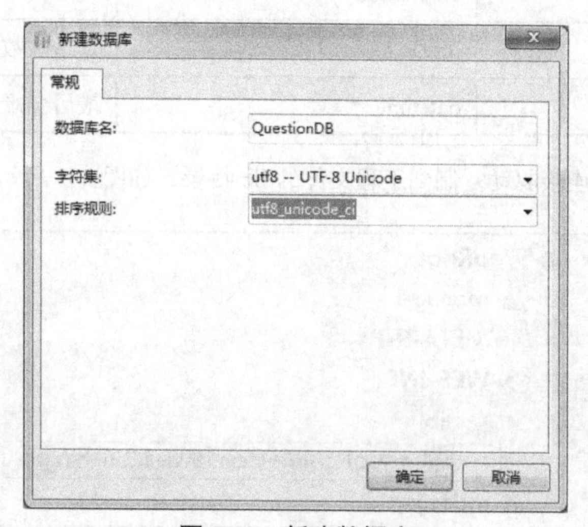

图 5-1　新建数据库

并在此数据库中新建一个关系表，表结构如图 5-2 和表 5-2 所示。

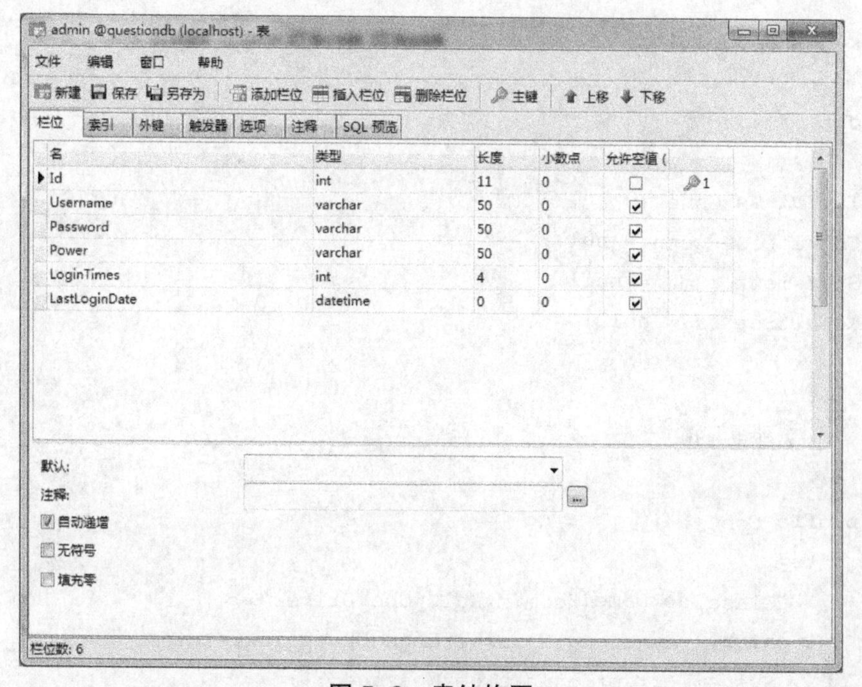

图 5-2　表结构图

表 5-2 新建表

字段名	字段类型	说明
Id	int	自动增长序号
Username	varchar(50)	用户名
Password	varchar(50)	密码
Power	varchar(50)	权限
LoginTimes	int	登录次数
LastLoginDate	datetime	最后登录时间

（4）将 MySQL 连接驱动复制到工程文件的 lib 目录，如图 5-3 所示。

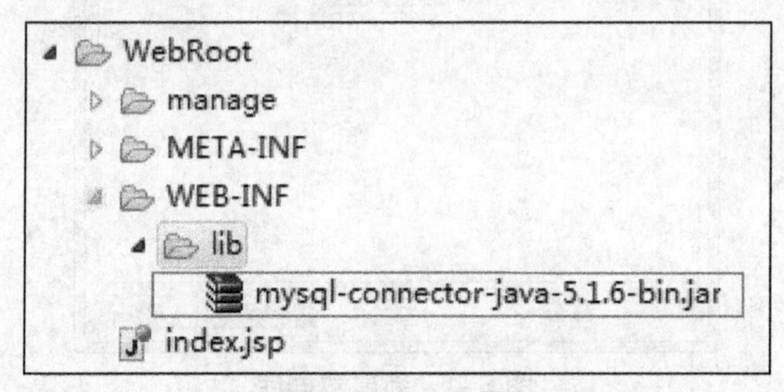

图 5-3 lib 目录

（5）在 Web project 中创建连接数据库的 JavaBean、com.db.Conn.java，核心代码如下：

```
package com.db;

import java.sql.*;

public class Conn {
   Connection conn = null;
   Statement stmt = null;
   ResultSet rs = null;

   /**
    * 加载驱动程序
    * */
   public Conn() {
      try {
         Class.forName("com.mysql.jdbc.Driver");
      } catch (java.lang.ClassNotFoundException e) {
         System.err.println(e.getMessage());
```

```
        }
    }

    /**
     * 执行查询操作: select
     * */
    public ResultSet executeQuery(String sql) {
        try {
            conn = DriverManager.getConnection("jdbc:mysql://localhost:3306/
questiondb?useUnicode=true&characterEncoding=UTF-8", "root", "123456");
// 根据安装 MySQL 的用户名与密码进行设置
            stmt = conn.createStatement();
            rs = stmt.executeQuery(sql);
        } catch (SQLException ex) {
            System.err.println(ex.getMessage());
        }
        return rs;
    }

    /**
     * 执行更新操作: insert、update、delete
     * */
    public int executeUpdate(String sql) {
        int result = 0;
        try {
            conn = DriverManager.getConnection("jdbc:mysql://localhost:3306/
questiondb?useUnicode=true&characterEncoding=UTF-8", "root", "123456");
// 根据安装 MySQL 的用户名与密码进行设置
            stmt = conn.createStatement(ResultSet.TYPE_SCROLL_INSENSITIVE,
ResultSet.CONCUR_READ_ONLY);
            result = stmt.executeUpdate(sql);
        } catch (SQLException ex) {
            result = 0;
        }
        return result;
    }

    /**
     * 关闭数据库连接
     * */
    public void close() {
```

```
        try {
            if (rs != null)
                rs.close();
        } catch (Exception e) {
            e.printStackTrace(System.err);
        }
        try {
            if (stmt != null)
                stmt.close();
        } catch (Exception e) {
                e.printStackTrace(System.err);
        }
        try {
            if (conn != null) {
                conn.close();
            }
        } catch (Exception e) {
            e.printStackTrace(System.err);
        }
    }
}
```

（6）在此工程中创建一个模型 com.model.AdminInfo.java，该模型只有一些属性及其getter 与 setter 方法的类，这些类没有业务逻辑。核心代码如下：

```
package com.model;
import java.util.Date;
public class AdminInfo {
    private int id;
    private String username;
    private String password;
    private String power;
    private int logintimes;
    private Date lastlogindate;
    public int getId() {
        return id;
    }
    public void setId(int id) {
        this.id = id;
    }
    public String getUsername() {
        return username;
```

```
    }
    public void setUsername(String username) {
        this.username = username;
    }
    public String getPassword() {
        return password;
    }
    public void setPassword(String password) {
        this.password = password;
    }
    public String getPower() {
        return power;
    }
    public void setPower(String power) {
        this.power = power;
    }
    public int getLogintimes() {
        return logintimes;
    }
    public void setLogintimes(int logintimes) {
        this.logintimes = logintimes;
    }
    public Date getLastlogindate() {
        return lastlogindate;
    }
    public void setLastlogindate(Date lastlogindate) {
        this.lastlogindate = lastlogindate;
    }

}
```

（7）在此工程中创建业务操作类 com.dal.Admin.java，核心代码如下：

```
package com.dal;

import java.sql.ResultSet;
import java.sql.SQLException;
import com.db.*;
import com.model.AdminInfo;

public class Admin {
    Conn conn = new Conn();
```

```
        /**
         * 判断当前登录用户是否存在
        */
        public boolean isExist(String username, String password) throws SQLException {
            boolean result = false;
            AdminInfo info = new AdminInfo();
             String sql = "select * from Admin where UserName='" + username
+ "' and Password='" + password + "'";
            // System.out.println(sql);
            ResultSet rs = conn.executeQuery(sql);
            if (rs.next()) {
                info.setUsername(rs.getString("UserName"));
                result = true;
            }
            conn.close();
            return result;
        }

    }
```

（8）实现 login_check.jsp，在此文件中调用业务操作类 com.dal.Admin.java。核心代码如下：

```
    <%@page language="java" import="java.util.*" pageEncoding="utf-8"%>
    <%@page import="com.dal.Admin"%>
    <%@page import="com.model.AdminInfo"%>
    <%@page import="com.common.Utility"%>

    <%
        if ("LoginAction".equals(request.getParameter("action"))) {
            String txtUsername = request.getParameter("txtUsername");
            String txtPassword = request.getParameter("txtPassword");
           //接下来从数据库中判断用户输入的用户名与密码是否正确
            Admin admin = new Admin();
            if (!admin.isExist(txtUsername, txtPassword)) {
                out.println(" <script>alert('用户名密码有误');window.location.
href('login.jsp');</script>");
            } else {
                Utility.writeCookie(response, "admin", txtUsername); //记录
登录状态，写入到 cookie 中
                response.sendRedirect("main.jsp");
            }
```

```
        } else {
            out.print("<script  language=\"javascript\">alert('非法访问！');
window.loaction.href='login.jsp';</script>");

        }
    %>
```

（9）编写 Cookie 读 / 写操作的通用类，命名为 Utility.java。核心代码如下：

```
    package com.common;
    import javax.servlet.http.Cookie;
    import javax.servlet.http.HttpServletRequest;
    import javax.servlet.http.HttpServletResponse;

    /**
     * 通用操作类
     */
    public class Utility {

        /**
         * 写入 cookie
         * @param response
         * @param key
         * @param value
         */
        public static void writeCookie(HttpServletResponse response, String
key, String value) {
            writeCookie(response, key, value, -1);
        }

        /**
         * 写入 cookie
         * @param response
         * @param key
         * @param value
         * @param expirse
         */
        public static void writeCookie(HttpServletResponse response, String
key, String value, int expirse) {
            Cookie newCookie = new Cookie(key, value);
            if (expirse > 0)
```

```
            expirse = expirse * 60;
        newCookie.setPath("/");
        newCookie.setMaxAge(expirse);
        response.addCookie(newCookie);
    }

    /**
     * 读取cookie值
     * @param request
     * @param key
     * @return
     */
    public static String readCookie(HttpServletRequest request, String key) {
        String value = "";
        Cookie[] ck = request.getCookies();
        if (ck == null)
            return "";
        for (Cookie c : ck) {
            if (c.getName().equals(key)) {
                value = c.getValue();
                break;
            }
        }
        return value;
    }
}
```

（10）退出登录，清空 Cookie。新建 logout.jsp，核心代码如下：

```
<%@page language="java" import="java.util.*" pageEncoding="utf-8"%>
<%@page import="com.common.Utility"%>
<%@page import="com.common.DataValidator"%>
<%
    String path = request.getContextPath();
    String basePath = request.getScheme() + "://" + request.getServerName()
+ ":" + request.getServerPort() + path + "/";
    String data = Utility.readCookie(request, "admin");//注意 key名称 admin
要与login_check.jsp中写入Cookie的key名称一样
    //判断data是否为空
    //类似这种数据验证，在本项目中许多地方仍然会用到，为提高代码的重用度，建议将其编
写为一个通用类，方便后面调用
    //这里编写一个类名为DataValidator的数据类型验证类，包名为com.common
```

```
    if (DataValidator.isNullOrEmpty(data)) {
        response.sendRedirect("login.jsp");
    } else {
        Utility.writeCookie(response, "admin", data, 0);
        response.sendRedirect("login.jsp");
    }
%>
```

数据验证类文件 DataValidator.java，核心代码如下：

```
package com.common;

/**
 * 数据验证类
 */
public class DataValidator {

    /**
     * 验证字符串是否为空 = "" or = null
     * @param input - 需要验证的字符串
     * @return true/false
     */
    public static boolean isNullOrEmpty(String input){
        return "".equals(input) || input == null;
    }

}
```

实验六

JSP+MySQL 添加操作实现后台录入问卷到数据库

实验目的

☑ 掌握 MySQL 数据库及 Navicat for MySQL 的使用。

☑ 熟练使用 SQL 语句进行数据的插入。

☑ 掌握 JSP 操作 MySQL 数据库的方法与步骤。

实验要求

（1）本实验要求在实验五的基础上进行。

（2）将本次实验的所有工程文件保留。

（3）实验结束后，按要求完成实验报告。

实验课时

2 课时。

实验准备

复习结构化查询语言中的 insert 语句。

实验内容

（1）打开 Navicat for MySQL 可视化操作软件。创建问题表 Question，表结构如图 6-1 所示。

questiondb
表
admin
question
视图
函数
事件
查询
报表
备份
sakila

名	类型	长度	小数点	不是 null	
Id	int	11	0	☑	1
Title	varchar	255	0	☐	
A	varchar	255	0	☐	
B	varchar	255	0	☐	
C	varchar	255	0	☐	
D	varchar	255	0	☐	
Answer	varchar	1	0	☐	
Grade	double	255	0	☐	

图 6-1　表结构

（2）创建该表的模型，目录结构如图 6-2 所示。

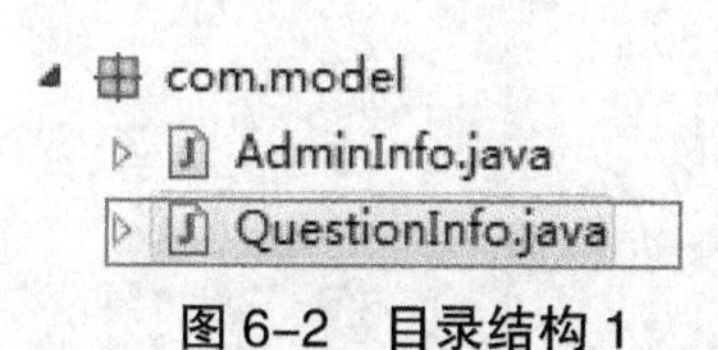

图 6-2　目录结构 1

核心代码如下：

```
package com.model;

public class QuestionInfo {
    private int id;
    private String title;
    private String a;
    private String b;
    private String c;
    private String d;
    private String answer;
    private double grade;

    public int getId() {
        return id;
    }

    public void setId(int id) {
        this.id = id;
    }

    public String getTitle() {
        return title;
    }
```

```
    public void setTitle(String title) {
        this.title = title;
    }

    public String getA() {
        return a;
    }

    public void setA(String a) {
        this.a = a;
    }

    public String getB() {
        return b;
    }

    public void setB(String b) {
        this.b = b;
    }

    public String getC() {
        return c;
    }

    public void setC(String c) {
        this.c = c;
    }

    public String getD() {
        return d;
    }

    public void setD(String d) {
        this.d = d;
    }

    public String getAnswer() {
        return answer;
    }

    public void setAnswer(String answer) {
```

```
        this.answer = answer;
    }

    public double getGrade() {
        return grade;
    }

    public void setGrade(double grade) {
        this.grade = grade;
    }

}
```

（3）创建针对该表的业务操作类 Question.java，如图 6-3 所示。该类中暂时只有一个插入方法 Insert()。核心代码如下：

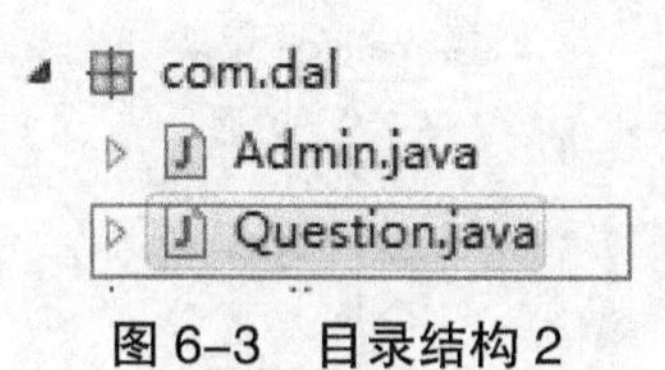

图 6-3　目录结构 2

```
package com.dal;

import java.sql.ResultSet;
import java.sql.SQLException;
import java.util.ArrayList;
import java.util.List;
import com.db.Conn;
import com.common.DataValidator;
import com.model.QuestionInfo;

public class Question {
    //创建连接数据库对象 conn
    Conn conn = new Conn();

    /**
     * 问题插入操作
     *
     * @param info
     * @return
     */
```

```
        public int insert(QuestionInfo info) {
            String sql = "insert into question(Title,A,B,C,D,Answer,Grade) values ";
            sql = sql + " ('" + info.getTitle() + "','" + info.getA() + "','"
+ info.getB() + "','" + info.getC() + "','" + info.getD() + "','" + info.
getAnswer() + "','" + info.getGrade() + "')";
            int result = 0;
            System.out.print(sql);
            result = conn.executeUpdate(sql);
            conn.close();
            return result;

        }
    }
```

（4）创建操作页面，录入问卷的问题，如图 6-4 和图 6-5 所示。

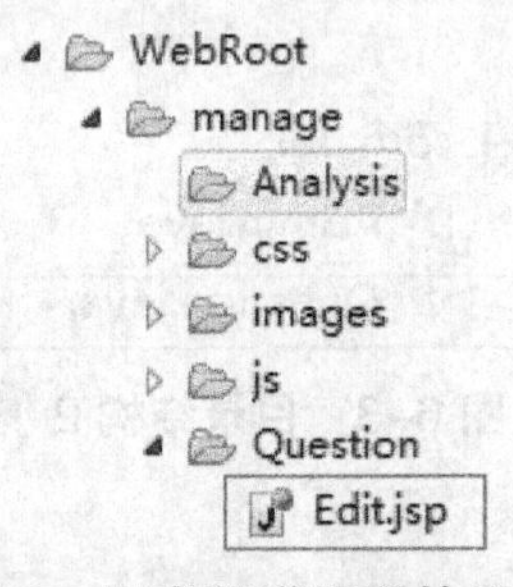

图 6-4　问卷新增问题的目录结构

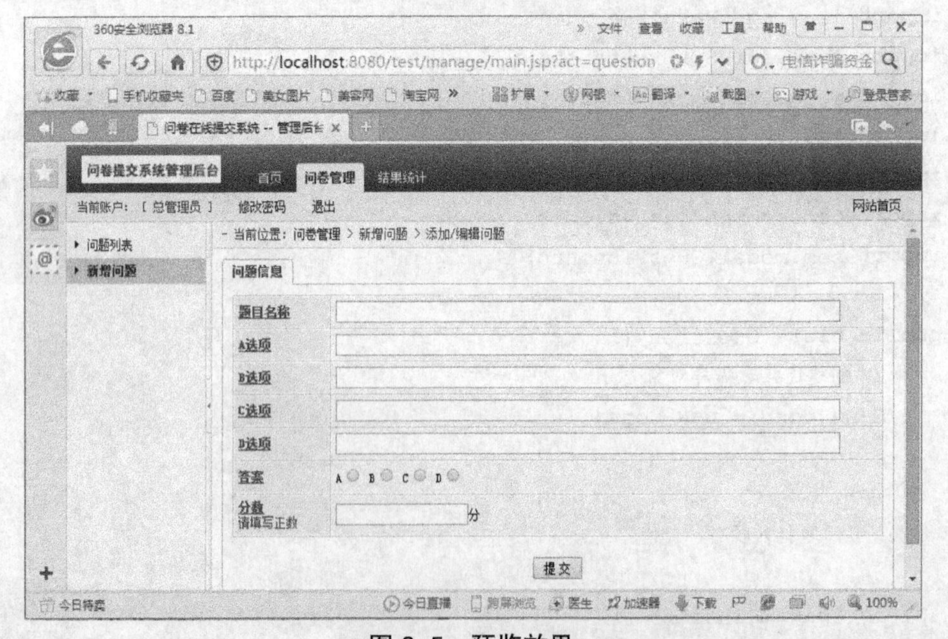

图 6-5　预览效果

（5）完善 Edit.jsp 页面，在此页中调用 com.dal.Question 类方法，实现问题的入库操作。

实验七

JSP+MySQL 编辑操作实现后台在线编辑问卷中的问题

实验目的

☑ 掌握 MySQL 数据库及 Navicat for MySQL 的使用。

☑ 熟练使用 SQL 语句进行数据的编辑。

☑ 掌握 JSP 操作 MySQL 数据库的方法与步骤。

实验要求

（1）本实验要求在实验六的基础上进行。

（2）将本次实验的所有工程文件保留。

（3）实验结束后，按要求完成实验报告。

实验课时

4 课时。

实验准备

复习结构化查询语言中的 update 语句。

实验内容

（1）新建一个 list.jsp 页面，保存在 Question 目录下，如图 7-1 所示。

图 7-1　目录结构

（2）从数据表 Question 中查询数据并显示在 list.jsp 页面，首先设计一个获取列表的类方法。核心代码如下：

```
/**
     * 获取问题列表
     *
     * @return
     * @throws SQLException
     */
    public List<QuestionInfo> getList(String keyword) throws SQLException {
        List<QuestionInfo> list = new ArrayList<QuestionInfo>();
        String sql = "select q.* from Question q  ";
        if (DataValidator.isNullOrEmpty(keyword)) {
            sql = sql + " order by id desc";
        } else {
            sql = sql + " where q.title like '%" + keyword + "%' order
by id desc";

        }
        ResultSet rs = conn.executeQuery(sql);

        while (rs.next()) {
            QuestionInfo info = new QuestionInfo();
            info.setId(rs.getInt("Id"));
            info.setTitle(rs.getString("Title"));
            info.setA(rs.getString("A"));
            info.setB(rs.getString("B"));
            info.setC(rs.getString("C"));
            info.setD(rs.getString("D"));
            info.setAnswer(rs.getString("Answer"));
```

```
            info.setGrade(rs.getDouble("Grade"));
            list.add(info);
        }
        conn.close();
        return list;
    }
```

（3）在 list.jsp 页面调用步骤（2）中的类方法，最终显示效果如图 7-2 所示。

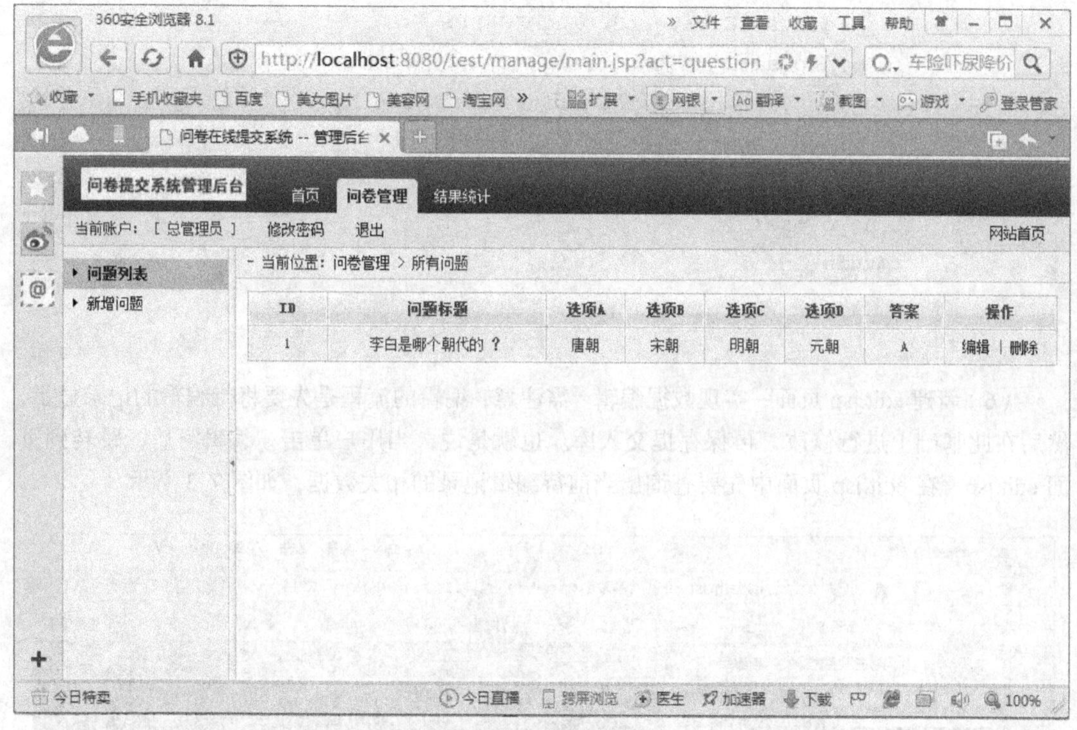

图 7-2　显示效果

（4）在 list.jsp 页面中，修改“编辑”链接内容，代码如下：

```
<a href="edit.jsp?action=Edit&id=?"> 编辑 </a>
```

思考一下上述代码中的 id=?，此处的问号应该如何处理，才能保证编辑的是当前记录。

（5）在 com.dal.Question 中新增加一个类方法。核心代码如下：

```
/**
     * 获得单条问题
     *
     * @param id
     * @return
     * @throws SQLException
     */
    public QuestionInfo getQuestionInfo(int id) throws SQLException {
```

```
        QuestionInfo info = new QuestionInfo();
        String sql = "select q.* from Question q where q.id=" + id + "";
        ResultSet rs = conn.executeQuery(sql);
        if (rs.next()) {
            info.setId(rs.getInt("Id"));
            info.setTitle(rs.getString("Title"));
            info.setA(rs.getString("A"));
            info.setB(rs.getString("B"));
            info.setC(rs.getString("C"));
            info.setD(rs.getString("D"));
            info.setAnswer(rs.getString("Answer"));
            info.setGrade(rs.getDouble("Grade"));
        }
        conn.close();
        return info;

    }
```

（6）新建 edit.jsp 页面，实现数据编辑。需注意，编辑的流程是先要将待编辑的记录显示，然后在此基础上进行修改，再保存提交入库。也就是说，当用户单击“编辑”后，跳转到页面 edit.jsp，在 edit.jsp 页面中先要查询出当前待编辑记录的相关数据，如图 7-3 所示。

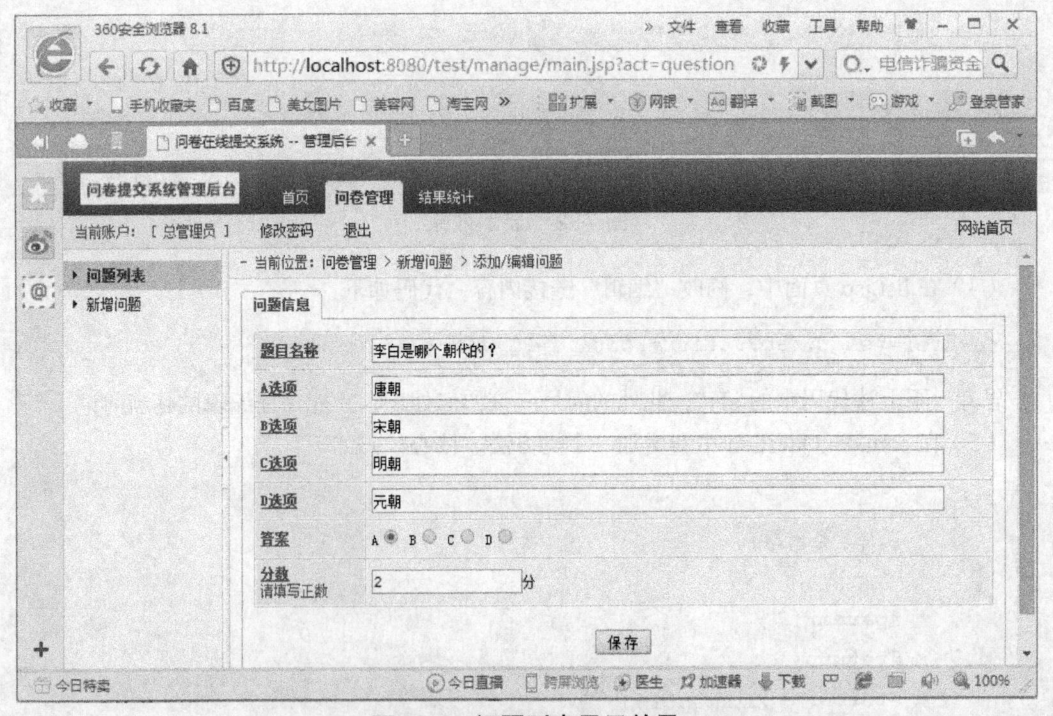

图 7-3　问题列表显示效果

核心代码如下：

```
<%@page language="java" import="java.util.*" pageEncoding="utf-8"%>
<%@include file="../is_login.jsp" %>
<%@page import="com.dal.Question"%>
<%@page import="com.model.QuestionInfo"%>
<%@page import="com.common.DataConverter"%>
<%
    String path = request.getContextPath();
    String basePath = request.getScheme() + "://" + request.getServerName()
+ ":" + request.getServerPort() + path + "/";
%>
<%
    request.setCharacterEncoding("utf-8");
    Question question = new Question();
    QuestionInfo info = new QuestionInfo();

    if ("Edit".equals(request.getParameter("action"))) {
            info = question.getQuestionInfo(DataConverter.toInt(request.
getParameter("hidId")));
        if (info == null) {
            out.print("找不到相关信息");
            return;
        }


    }
%>
<!DOCTYPE HTML PUBLIC "-//W3C//DTD HTML 4.01 Transitional//EN">
<html>
<head>

<title>问卷编辑</title>

<meta http-equiv="pragma" content="no-cache">
<meta http-equiv="cache-control" content="no-cache">
<meta http-equiv="expires" content="0">
<meta http-equiv="keywords" content="keyword1,keyword2,keyword3">
<meta http-equiv="description" content="This is my page">
<link rel="stylesheet" type="text/css" href="../css/admin_common.css" />


</head>
```

```
<body>
    <div class="path">
        - 当前位置：<a href="/manage/" target="_top">问卷管理</a> &gt; 新增问题 &gt; 添加/编辑问题
    </div>
    <div class="body">
      <form name="form1" method="post" action="edit.jsp?action=Edit">
          <div id="tab">
            <ul>
              <li class="actived"><span>问题信息</span></li>

            </ul>
          <div class="c"></div>
        </div>
        <div id="tab_content">
        <%
        QuestionInfo qInfo=question.getQuestionInfo(DataConverter.toInt(request.getParameter("id")));
        if (qInfo==null)
        {
            out.print("找不到相关信息");
            return;
        }
        %>
            <table id="tab_content_0" width="100%" border="0" cellspacing= "1" cellpadding="0" class="tableform">
            <tr>
            <td width="15%" class="item"><label>题目名称</label></td>
            <td width="85%" class="input"><input type="text" name="txtTitle" style="width:500px;" value="<%=qInfo.getTitle()%>"></td>
            </tr>
            <tr>
              <td class="item"><label>A 选项</label></td>
              <td class="input"><input type="text" name="txtA" style="width:500px;" value="<%=qInfo.getA()%>"></td>
            </tr>
            <tr>
              <td class="item"><label>B 选项</label></td>
              <td class="input"><input type="text" name="txtB" style="width:500px;" value="<%=qInfo.getB()%>"></td>
```

```
            </tr>
            <tr>
              <td class="item"><label>C 选项 </label></td>
              <td class="input"><input type="text" name="txtC" style="width:
500px;" value="<%=qInfo.getC()%>"> </td>
            </tr>
            <tr>
              <td class="item"><label>D 选项 </label></td>
              <td class="input"><input type="text" name="txtD" style="width:
500px;" value="<%=qInfo.getD()%>"></td>
            </tr>
            <tr>
              <td class="item"><label> 答案 </label></td>
              <td class="input">A<input type="radio" value="a" name="radioAnswer"
<%if ("a".equals(qInfo.getAnswer())) out.print("checked"); %>> B<input
type="radio" value="b" name="radioAnswer" <%if ("b".equals(qInfo.getAnswer()))
out.print("checked"); %>> C<input type="radio" value="c" name="radioAnswer"
<%if ("c".equals(qInfo.getAnswer())) out.print("checked"); %>> D<input
type="radio" value="d" name="radioAnswer" <%if ("d".equals(qInfo.getAnswer()))
out.print("checked"); %>></td>
            </tr>
            <tr>
              <td class="item"><label> 分数 </label><i> 请填写正数 </i></td>
              <td class="input"><input type="text" name="txtGrade" value="<%=
qInfo.getGrade()%>"> 分 </td>
            </tr>
            </table>
            <div id="btnMain" class="operations">
              <input type="submit" value=" 保 存 " />
              <input type="hidden" value="<%=qInfo.getId()%>" name= "hidId" />
          </div>
          </div>
      </form>
    </div>
  </body>
  </html>
```

（7）在类 com.dal.Question 中新建一个类方法 update()。核心代码如下：

```
    /**
     * 问题修改操作
     * @param info
```

```
        * @return
        */
       public int update(QuestionInfo info) {
           String sql = "update question set" + " Title='" + info.getTitle()
+ "',A='" + info.getA() + "',B='" + info.getB() + "',C='" + info.getC() + "',D='"
+ info.getD() + "',Answer='" + info.getAnswer() + "',Grade='" + info.getGrade() +
"' where id=" + info.getId() + "";
           int result = 0;
           System.out.println(sql);
           result = conn.executeUpdate(sql);
           conn.close();
           return result;
       }
```

修改 edit.jsp 页面，单击“保存”按钮后的代码如下：

```
   <%@ page language="java" import="java.util.*" pageEncoding="utf-8"%>
   <%@include file="../is_login.jsp" %>
   <%@page import="com.dal.Question"%>
   <%@page import="com.model.QuestionInfo"%>
   <%@page import="com.common.DataConverter"%>
   <%
       String path = request.getContextPath();
       String basePath = request.getScheme() + "://" + request.getServerName() +
":" + request.getServerPort() + path + "/";
   %>
   <%
       request.setCharacterEncoding("utf-8");
       Question question = new Question();
       QuestionInfo info = new QuestionInfo();

       if ("Edit".equals(request.getParameter("action"))) {
            info = question.getQuestionInfo(DataConverter.toInt(request.
getParameter("hidId")));
           if (info == null) {
               out.print("找不到相关信息");
               return;
           }
           info.setTitle(request.getParameter("txtTitle"));
           info.setA(request.getParameter("txtA"));
           info.setB(request.getParameter("txtB"));
           info.setC(request.getParameter("txtC"));
```

```
            info.setD(request.getParameter("txtD"));
            info.setAnswer(request.getParameter("radioAnswer"));
            info.setGrade(DataConverter.toDouble(request.getParameter("txtGrade")));
            int result = question.update(info); //update() 方法实现对象的修改
            if (result == 0) {
              out.print("<script>alert(\"编辑失败！\");window.location.href='list.jsp';
</script>");
            } else {
              out.print("<script>alert(\"编辑成功！\");window.location.href='list.jsp';
</script>");
            }

        }
    %>
    <!DOCTYPE HTML PUBLIC "-//W3C//DTD HTML 4.01 Transitional//EN">
    <html>
    <head>

    <title> 问卷编辑 </title>

    <meta http-equiv="pragma" content="no-cache">
    <meta http-equiv="cache-control" content="no-cache">
    <meta http-equiv="expires" content="0">
    <meta http-equiv="keywords" content="keyword1,keyword2,keyword3">
    <meta http-equiv="description" content="This is my page">
    <link rel="stylesheet" type="text/css" href="../css/admin_common.css" />


    </head>

    <body>
        <div class="path">
            - 当前位置：<a href="/manage/" target="_top"> 问卷管理 </a> &gt; 新
增问题 &gt; 添加 / 编辑问题
        </div>
        <div class="body">
          <form name="form1" method="post" action="edit.jsp?action=Edit">
              <div id="tab">
              <ul>
              <li class="actived"><span> 问题信息 </span></li>
```

```
            </ul>
            <div class="c"></div>
            </div>
        <div id="tab_content">
        <%
            QuestionInfo qInfo=question.getQuestionInfo(DataConverter.toInt
(request.getParameter("id")));
            if (qInfo==null)
            {
                out.print("找不到相关信息");
                return;
            }
            %>
            <table id="tab_content_0" width="100%" border="0" cellspacing= "1"
cellpadding="0" class="tableform">
            <tr>
                <td width="15%" class="item"><label>题目名称</label></td>
                <td width="85%" class="input"><input type="text" name="txtTitle"
style="width:500px;" value="<%=qInfo.getTitle()%>"></td>
            </tr>
            <tr>
                <td class="item"><label>A选项</label></td>
                <td class="input"><input type="text" name="txtA" style="width:
500px;" value="<%=qInfo.getA()%>"></td>
            </tr>
            <tr>
                <td class="item"><label>B选项</label></td>
                <td class="input"><input type="text" name="txtB" style="width:
500px;" value="<%=qInfo.getB()%>"></td>
            </tr>
            <tr>
                <td class="item"><label>C选项</label></td>
                <td class="input"><input type="text" name="txtC" style="width:
500px;" value="<%=qInfo.getC()%>"> </td>
            </tr>
            <tr>
                <td class="item"><label>D选项</label></td>
                <td class="input"><input type="text" name="txtD" style="width:
500px;" value="<%=qInfo.getD()%>"></td>
            </tr>
            <tr>
```

```
                <td class="item"><label>答案</label></td>
                <td class="input">A<input type="radio" value="a" name="radioAnswer"
<%if ("a".equals(qInfo.getAnswer())) out.print("checked"); %>> B<input
type="radio" value="b" name="radioAnswer" <%if ("b".equals(qInfo.getAnswer()))
out.print("checked"); %>> C<input type="radio" value="c" name="radioAnswer"
<%if ("c".equals(qInfo.getAnswer())) out.print("checked"); %>> D<input
type="radio" value="d" name="radioAnswer" <%if ("d".equals(qInfo.getAnswer()))
out.print("checked"); %>></td>
            </tr>
            <tr>
                <td class="item"><label>分数</label><i>请填写正数</i></td>
                <td class="input"><input type="text" name="txtGrade" value=
"<%=qInfo.getGrade()%>">分</td>
            </tr>
        </table>

        <div id="btnMain" class="operations">
            <input type="submit" value=" 保 存 " />
            <input type="hidden" value="<%=qInfo.getId()%>" name="hidId" />
        </div>
        </div>
        </form>
        </div>

    </body>
    </html>
```

实验八

JSP+MySQL 删除操作
实现后台在线删除问卷中的问题

实验目的

☑ 掌握 MySQL 数据库及 Navicat for MySQL 的使用。

☑ 熟练使用 SQL 语句进行数据的删除。

☑ 掌握 JSP 操作 MySQL 数据库的方法与步骤。

实验要求

（1）本实验要求在实验七的基础上进行。

（2）将本次实验的所有工程文件保留。

（3）实验结束后，按要求完成实验报告。

实验课时

2 课时。

实验准备

复习结构化查询语言中的 delete 语句。

实验内容

（1）新建一个 delete.jsp 页面，保存在 Question 目录下，如图 8-1 所示。

```
◢ WebRoot
  ◢ manage
    ▷ Analysis
    ▷ css
    ▷ images
    ▷ js
    ◢ Question
        add.jsp
        delete.jsp
        edit.jsp
```

图 8-1　目录结构

（2）在 com.dal.Question 类中新增加一个删除的方法，代码如下：

```
/**
```

```
     * 问题删除操作
     *
     * @param id
     * @return
     */
    public int delete(int id) {
        String sql = "delete from question where id =" + id + "";
        int result = 0;
        result = conn.executeUpdate(sql);
        conn.close();
        return result;
    }
```

（3）在 list.jsp 页面中，修改“删除”链接内容，代码如下：

```
<a href="<%=basePath%>manage/Question/delete.jsp?action=Delete&id=<%=
info.getId()%>" onclick="return confirm('确定要删除该记录吗？');">删除</a>
```

思考一下上述代码中的 action=Delete 参数存在的意义。

（4）delete.jsp 页面调用步骤（2）中的类方法，代码如下：

```
<%@page language="java" import="java.util.*" pageEncoding="utf-8"%>
<%@page import="com.dal.Question"%>
<%@page import="com.model.QuestionInfo"%>
<%@page import="com.common.DataConverter"%>
<%
if ("Delete".equals(request.getParameter("action"))) {
    Question question = new Question();
    int result = 0;
    result = question.delete(DataConverter.toInt(request.getParameter("id")));
    if (result == 0) {
        out.print("<script>alert('删除失败!');window.loaction.href='list.jsp';
</script>");
    } else {
        out.print("<script>alert('删除成功!');window.loaction.href='list.jsp';
</script>");
    }
} else {
    out.print("<script>alert('无效的URL!');window.loaction.href='list.jsp';
</script>");
}%>
```

实验九
前台问卷收集

实验目的

☑ 掌握 MySQL 数据库及 Navicat for MySQL 的使用。

☑ 掌握表单控件动态生成并批量获取数值的方法。

实验要求

（1）本实验要求在实验八的基础上进行。

（2）将本次实验的所有工程文件保留。

（3）实验结束后，按要求完成实验报告。

实验课时

2 课时。

实验准备

如果想制作更为专业的统计图标，可以自行下载绘图插件 jqPlot。

实验内容

（1）新建一个表，命名为 answer，表结构如图 9-1 所示。

字段 | 索引 | 外键 | 触发器 | 选项 | 注释 | SQL 预览

名	类型	长度	小数点	不是 null
Id	int	11	0	☑
QuestionId	int	11	0	☐
Choice	varchar	2	0	☐

图 9-1　表结构

（2）创建实体模型 com.model.AnswerInfo 类，代码如下：

```
package com.model;

public class AnswerInfo {
    private int id;
    private int QuestionId;
    private String Choice;


    public int getId() {
        return id;
    }

    public void setId(int id) {
        this.id = id;
    }

    public int getQuestionId() {
        return QuestionId;
   }

    public void setQuestionId(int questionId) {
        QuestionId = questionId;
    }

    public String getChoice() {
        return Choice;
    }

    public void setChoice(String choice) {
        Choice = choice;
    }
}
```

（3）创建数据操作类 com.dal.Answer 类，代码如下：

```
package com.dal;
import java.sql.ResultSet;
import java.sql.SQLException;
import java.util.ArrayList;
import java.util.List;
import com.db.Conn;
```

```
import com.common.DataValidator;
import com.model.AnswerInfo;

public class Answer {
    // 创建连接数据库对象 conn
    Conn conn = new Conn();

    /**
     * 问题插入操作
     *
     * @param info
     * @return
     */
    public int insert(AnswerInfo info) {
        String sql = "insert into Answer(QuestionId,Choice) values ";
        sql = sql + " ('" + info.getQuestionId() + "','" + info.getChoice() + "')";
        int result = 0;
        System.out.println(sql);
        result = conn.executeUpdate(sql);
        conn.close();
        return result;

    }

}
```

（4）编辑根目录下的 index.jsp 页面，代码如下：

```
<div style=" margin:0px auto; line-height:50px;text-align:center;">问卷
在线提交系统</div>
    <form name="" action="./save.jsp?action=save" method="post">
        <table width="80%" border="0" align="center" cellpadding="0" cellspacing="0">
            <%
                Question question = new Question();
                List<QuestionInfo> list = question.getList("");
                int i = 1;
                for (QuestionInfo info : list) {
            %>
            <tr>
                <td height="30" bgcolor="#94A6DA" style="color:#ffffff;"><input
type="hidden" name="hidQuestionId" value="<%=info.getId()%>"><%=i%>、 <%=info.
getTitle()%></td>
```

```
            </tr>
            <tr>
                <td height="30">A.<input type="radio" name="choice<%=info.getId()%>" value="a"><%=info.getA()%> B.<input type="radio" name="choice<%=info.getId()%>" value="b"><%=info.getB()%> C.<input type="radio" name="choice<%=info.getId()%>" value="c"><%=info.getC()%> D.<input type="radio" name="choice<%=info.getId()%>" value="d"><%=info.getD()%>
                </td>
            </tr>
            <%
                i++;
                }
            %>
        </table>
        <div style=" margin:0px auto; line-height:50px;text-align:center;">
            <input type="submit" name="button" id="button" value="提交" onclick="return confirm('确定要提交吗？一经提交将无法修改！');" />
        </div>
    </form>
```

（5）在根目录下新建一个 save.jsp 页面，代码如下：

```
<%@page language="java" import="java.util.*" pageEncoding="utf-8"%>
<%@page import="com.model.AnswerInfo"%>
<%@page import="com.common.DataConverter"%>
<%@page import="com.dal.Answer"%>
<%
    request.setCharacterEncoding("utf-8");
    Answer question = new Answer();
    AnswerInfo info = new AnswerInfo();
    if ("save".equals(request.getParameter("action"))) {

        String choice = "";
        int questionId = 0;
        String[] questionIdList = request.getParameterValues("hidQuestionId");
        for (int i = 0; i <= questionIdList.length - 1; i++) {
            choice = request.getParameter("choice" + questionIdList[i]);
            //获取每一题的选项
            questionId = DataConverter.toInt(questionIdList[i]);
            //获取每一题的 ID
            //out.print(questionIdList[i] + "---" + choice + ",");
```

```
                //输出各问题的编号与选择的结果
                info.setChoice(choice);
                info.setQuestionId(questionId);
                int result = question.insert(info);
                if (result == 0) {
                    out.print("<script>alert('提交失败！ ');window.loaction.
href='index.jsp';</script>");
                } else {
                    out.print("<script>alert('提交成功！ ');window.loaction.
href='result.jsp';</script>");
                }
            }
        }
    %>
```

实验十
统计分析柱状图的设计与实现

实验目的

☑ 掌握 MySQL 数据库及 Navicat for MySQL 的使用。
☑ 熟练使用 SQL 语句中常见函数的使用。
☑ 掌握 Web 应用开发中分析统计报表的制作原理。

实验要求

（1）本实验要求在实验九的基础上进行。
（2）将本次实验的所有工程文件保留。
（3）实验结束后，按要求完成实验报告。

实验课时

2 课时。

实验准备

如果想制作更为专业的统计图标，可以自行下载绘图插件 jqPlot。

实验内容

（1）在 com.dal.Answer 类中新加一个统计分析的方法 getChoiceCount，代码如下：

```
public int getChoiceCount(int questionId,String choice) throws SQLException {
    String sql = "select count(1) as choicecount from Answer where
QuestionId='" + questionId + "' and Choice='"+choice+"'";
    int result = 0;
    System.out.println(sql);
    ResultSet rs = conn.executeQuery(sql);
```

```
        if (rs.next()) {
            result = rs.getInt("choicecount");
        }
        conn.close();
        return result;
    }
```

（2）新建 result.jsp 页面，以显示统计结果，目录结构如图 10-1 所示。

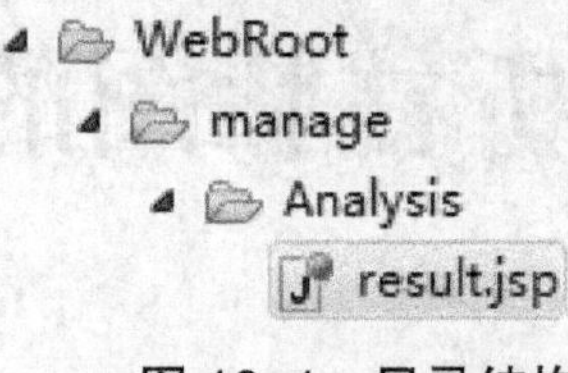

图 10-1　目录结构

核心代码如下：

```
    <div id="tab_content">
    <table id="tab_content_0" width="100%" border="0" cellspacing="1"
cellpadding="0" class="tableform">
    <%
        Answer answer = new Answer();
        Question question = new Question();
        List<QuestionInfo> list = question.getList("");
        int i = 1;
        for (QuestionInfo info : list) {
        %>
        <tr>
            <td width="25%" class="item"><label>题目 <%=i%></label></td>
            <td width="75%" class="input"><%=info.getTitle()%></td>
            </tr>
        <tr>
            <td class="item"></td>
            <td class="input" style="text-align: left;">

            <table height="136" border="0" align="left" cellpadding="0"
cellspacing="0">
                <tr>
                    <td  height="34">A.<%=info.getA()%>
                    </td>
                    <td ><div style="height:15px; width:<% =answer.getChoiceCount
(info.getId(),"a")*10%>px; background-color:blue;float: left;"></div></td>
                    </tr>
                <tr>
```

```
                    <td height="34">B.<%=info.getB()%>
                    </td>
                    <td><div style="height:15px; width:<%=answer.getChoiceCount
(info.getId(),"b")*10%>px; background-color:#CC3300;float:left;"></div></td>
                </tr>
                <tr>
                    <td height="34">C.<%=info.getC()%>
                    </td>
                    <td><div style="height:15px; width:<%=answer.getChoiceCount
(info.getId(),"c")*10%>px; background-color:#993333;float:left;"></div></td>
                </tr>
                <tr>
                    <td height="34">D.<%=info.getD()%></td>
                    <td><div style="height:15px; width:<%=answer.getChoiceCount
(info.getId(),"d")*10%>px; background-color:#663399;float:left;"></div></td>
                </tr>
                </table>
            </tr>
            <%
              i++;
              }
            %>
        </table>
        </div>
```

预览效果如图 10-2 所示：

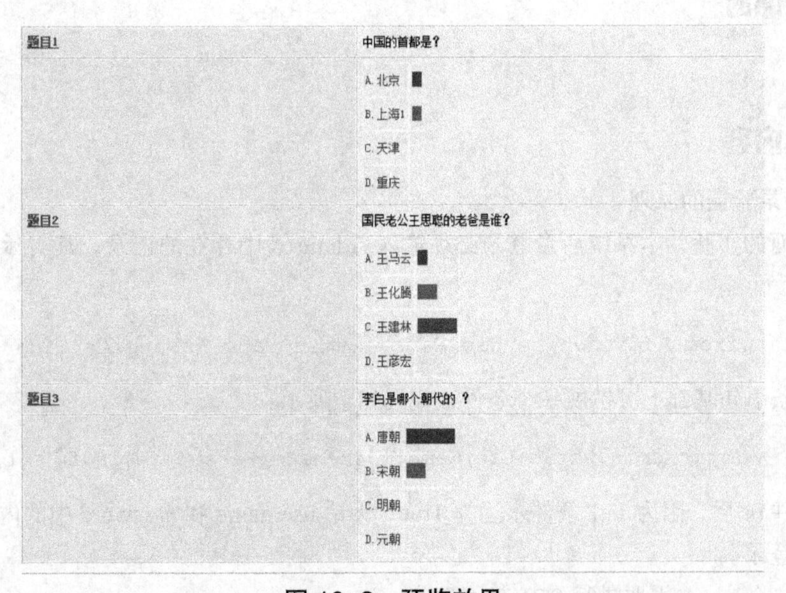

图 10-2　预览效果

实验十一

系统后台安全策略设计

实验目的

☑ 掌握常见的网站后台攻击方法与防御方案。

☑ 掌握常见的数据加密方法。

实验要求

（1）本实验要求在实验十的基础上进行。

（2）将本次实验的所有工程文件保留。

（3）实验结束后，按要求完成实验报告。

实验课时

2 课时。

实验内容

（1）登录验证的原理。

通过前面的实验，本系统后台登录必须输入 Admin 表中存在的记录。其实际验证的 SQL 语句为：

```
select * from Admin where UserName='admin' and Password='1234'
```

在上一条语句基础上可扩展一个条件，代码如下所示：

```
select * from Admin where UserName='34n' and Password='1sdf4'  or  '1=1'
```

添加条件 or 后，因为 1=1 条件永远为 True，无论 username 和 password 中的内容是什么，均可以正常登录。

这便是一个 bug，需要进行 SQL 注入防御。

（2）验证 SQL 注入登录。

打开系统的登录界面，输入任意用户名，密码输入：1' or '1=1，观察运行结果。

（3）如何打好这个补丁。在获取每个表单元素后，过滤获取的内容是否有 SQL 关键字。核心代码如下：

```
/**
     * 过滤 SQL 关键字，对于参数化查询中不需要过滤，只有在拼接 SQL 语句时用
     * @param input - 需要过滤的字符串
     * @return 过滤后的字符串
     */
    public static String filterSqlKeyword(String input){
        if(DataValidator.isNullOrEmpty(input))
            return input;
        boolean flag = false;
        String[] array = new String[] {
                "select", "update", "insert", "delete", "declare", "@",
"exec", "dbcc", "alter", "drop", "create", "backup", "if", "else", "end", "and",
"or", "add", "set", "open", "close", "use", "begin", "retun", "as", "go",
"exists", "kill", "chr"
            };
        for(int i = 0; i<array.length;i++){
            if(input.contains(array[i])){
                input = input.replace(array[i], "");
                flag = true;
            }
        }
        if(flag){
            return filterSqlKeyword(input);
        }
        return input;
    }
```

（4）后台管理员密码加密。

为提高系统的安全性，密码在任何系统中，均是以密文的形式保存。例如，一个字符串“abcd”，加密后的密文变成“e2fc714c4727ee9395f324cd2e7f331f”。将密文保存到数据库，即使数据库中数据发生泄露，也无法在短时间内将密文转换为明文。一定程度上保护了数据安全。

关于数据加密的算法很多，下面提供一个 MD5 算法加密的类。

```
package com.common;

import java.io.UnsupportedEncodingException;
```

```java
import java.security.MessageDigest;
import java.security.NoSuchAlgorithmException;

/**
 * MD5 加密类
 */
public class MD5 {
    private static MessageDigest digest = null;

    /**
     * 加密类，此方法默认为 16 位加密
     *
     * @param data
     * @return
     */
    public synchronized static final String Encrypt(String data) {
        return Encrypt(data, 16);
    }

    /**
     * 加密类，此方法可以手动设置加密位数
     *
     * @param data
     * @param len
     * @return
     */
    public synchronized static final String Encrypt(String data, int len) {
        if (digest == null) {
            try {
                digest = MessageDigest.getInstance("MD5");
            } catch (NoSuchAlgorithmException e) {
                e.printStackTrace();
            }
        }
        if (len != 16 && len != 32)
            len = 16;
        try {
            digest.update(data.getBytes("UTF-8"));
        } catch (UnsupportedEncodingException e) {
        }
        String s = encodeHex(digest.digest());
```

```
        if (len == 16) {
            return s.substring(8, 24);
        }
        return s;
    }

    private static final String encodeHex(byte[] bytes) {
        int i;
        StringBuffer buf = new StringBuffer(bytes.length * 2);
        for (i = 0; i < bytes.length; i++) {
            if (((int) bytes[i] & 0xff) < 0x10) {
                buf.append("0");
            }
            buf.append(Long.toString((int) bytes[i] & 0xff, 16));
        }
        return buf.toString();
    }
}
```

（5）后台所有页面增加身份安全验证判断。

通过前面的实验，系统后台入口是 login.jsp 页面，输入正确的口令与密码才能进入系统。但是如果在浏览器中直接输入后台页面的 URL 地址，将绕过这个登录入口，这时登录入口形同虚设。为解决此问题，将设计一个公共文件 is_login.jsp，目录结构如图 11-1 所示，此文件在后台每个页面都要引用。

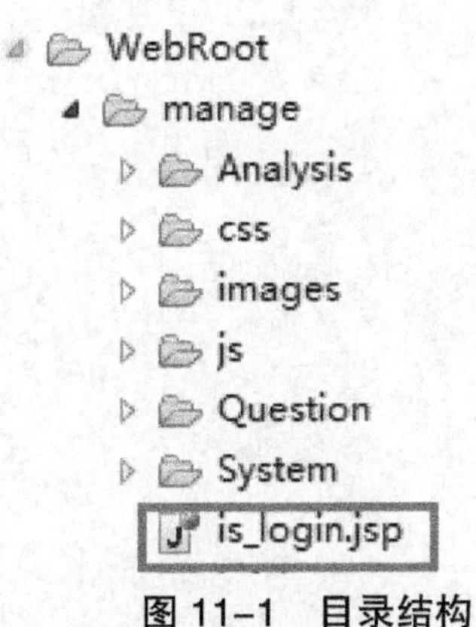

图 11-1　目录结构

此文件主要判断 Cookie 是否为空，如果是正常登录的，会写入一个 Cookie 值，如果没有正常登录，此 Cookie 为空，从而解决登录被绕开的问题。

```
<%@page language="java" import="java.util.*" pageEncoding="utf-8"%>
<%@page import="com.common.Utility"%>
<%@page import="com.common.DataValidator"%>
```

```
<%
    String loginPath = request.getScheme() + "://"
            + request.getServerName() + ":" + request.getServerPort()
            + request.getContextPath() + "/";
// 获取 Cookie 信息
String data = Utility.readCookie(request, "admin");

if(DataValidator.isNullOrEmpty(data)){
    response.sendRedirect(loginPath+"manage/login.jsp");
    return;
}
%>
```

在其他页面引入 is_login.jsp 文件。例如，Question/add.jsp 页面，需添加图 11-2 所示的代码，其他类似。

```
1 <%@ page language="java" import="java.util.*" pageEncoding="utf-8"%>
2 <%@include file="../is_login.jsp" %>
3 <%@ page import="com.model.QuestionInfo"%>
4 <%@ page import="com.dal.Question"%>
5 <%@ page import="com.common.DataConverter"%>
6 <%
```

图 11-2 其他页面引入 is_login.jsp 文件的代码

附录
MyEclipse 优化设置

1. 代码提醒功能

（1）Java 代码提醒

Java 代码提醒设置如图 1 和图 2 所示。

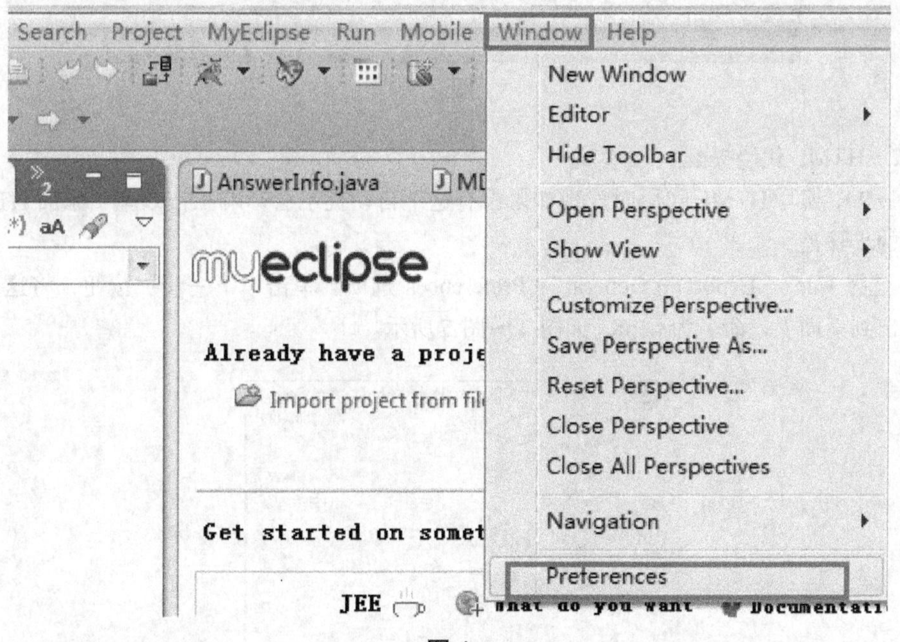

图 1

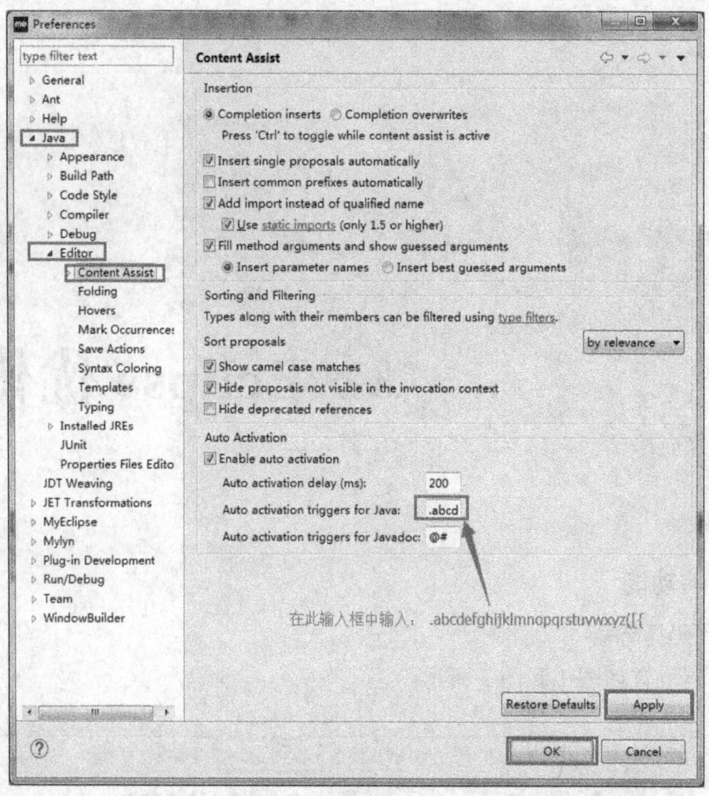

图 2

（2）HTML 代码提醒

在 JSP 中编写 HTML 代码时发现如果没有这个自动提示会影响工作效率。添加 HTML 代码提示的步骤是：

① 选择 File → Export → General → Preferences 选项。单击“下一步”按钮，将这个配置文件导出到桌面上，命名为 t.epf，如图 3 ～图 5 所示。

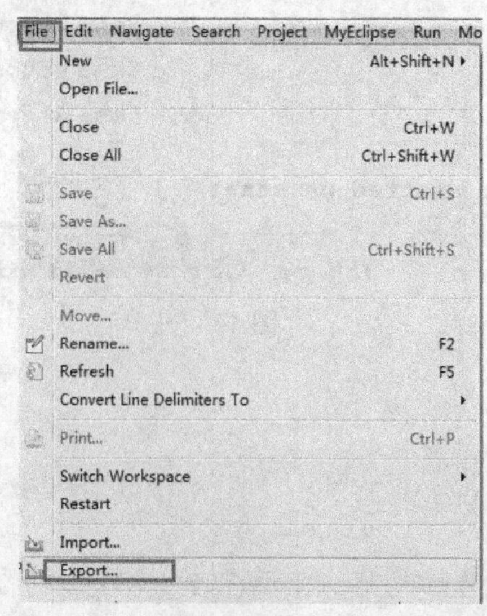

图 3

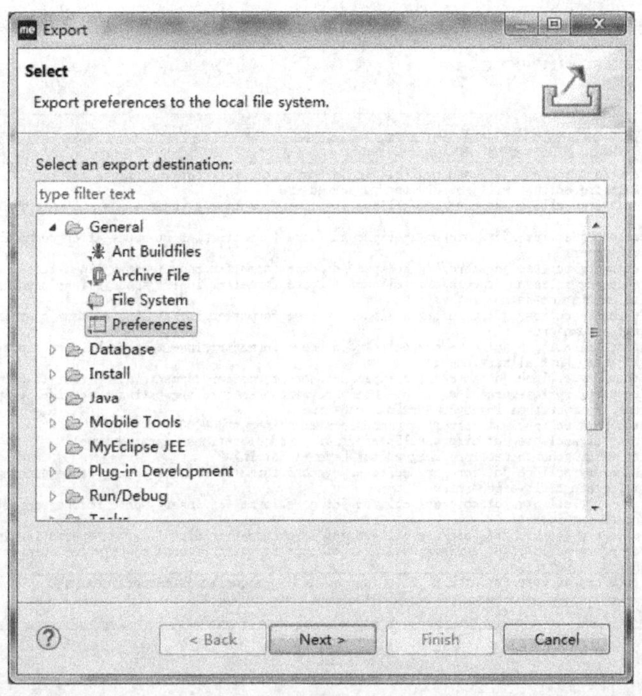

图 4

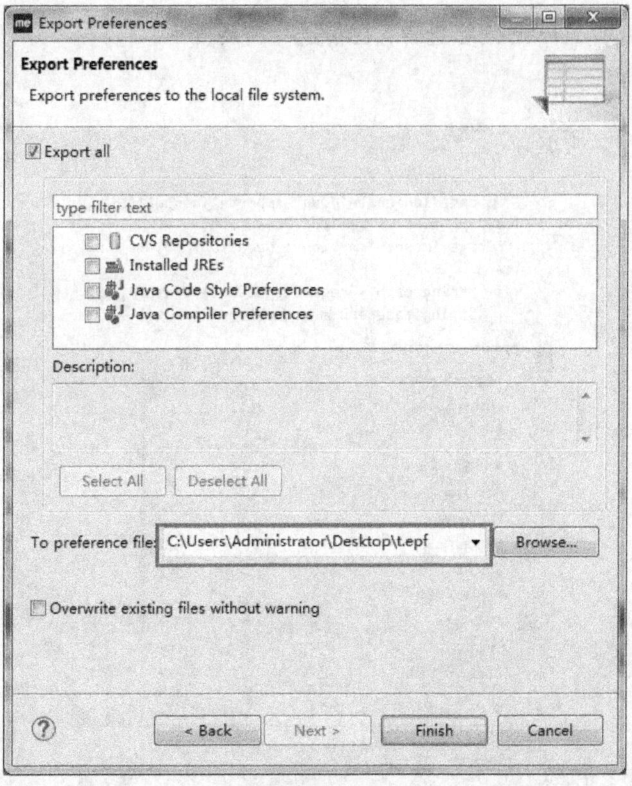

图 5

② 使用文本编辑器打开桌面上的文件，在末尾添加如下代码后保存，如图 6 所示。

```
    /instance/org.eclipse.wst.html.ui/autoProposeCode=<\=
abcdefghijklmnopqrstuvwxyz\:
```

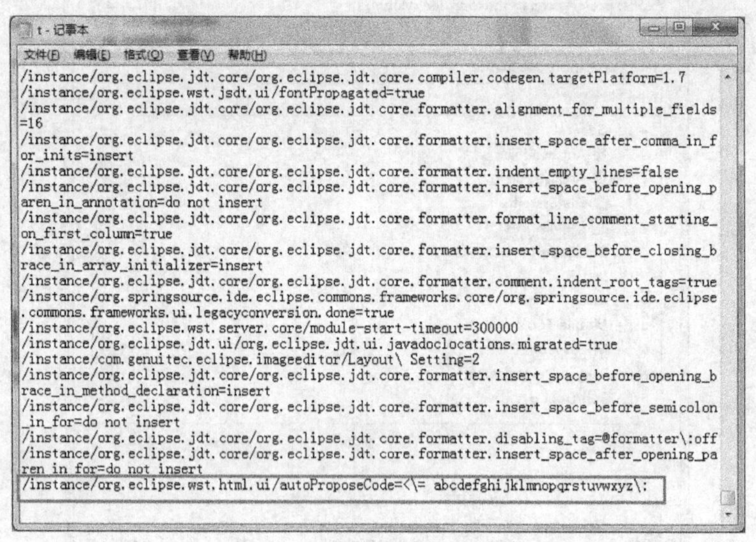

图 6

③ 将此配置文件导入，选择 File → Import → General → Preferences 选项，至此完成第一步的逆操作。

现在打开 JSP 或者 HTML 内容查看，已能进行自动提示。

2. 显示行号

行号的显示如图 7 所示。

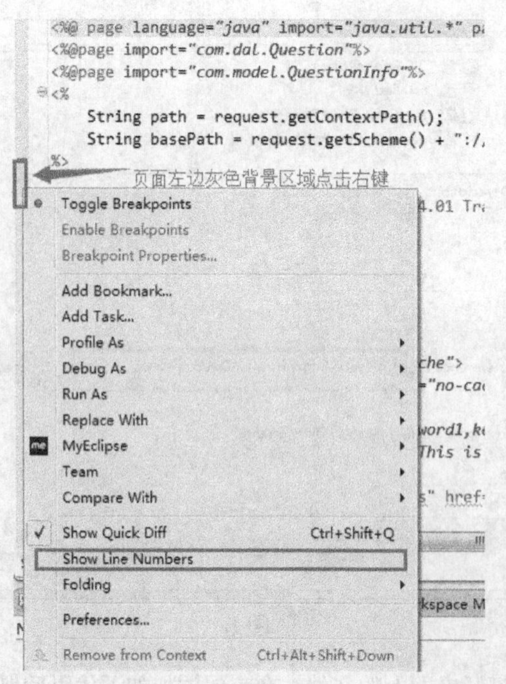

图 7

3. 提高 MyEclipse 每行代码的容量

提高 MyEclipse 每行代码的容量，防止代码自动换行，提高代码可读性。设置步骤如下：

```
Window → Preferences → Java → Code Style → Formatter → NEW
```

图 8 ～图 10 所示为各步骤截图。

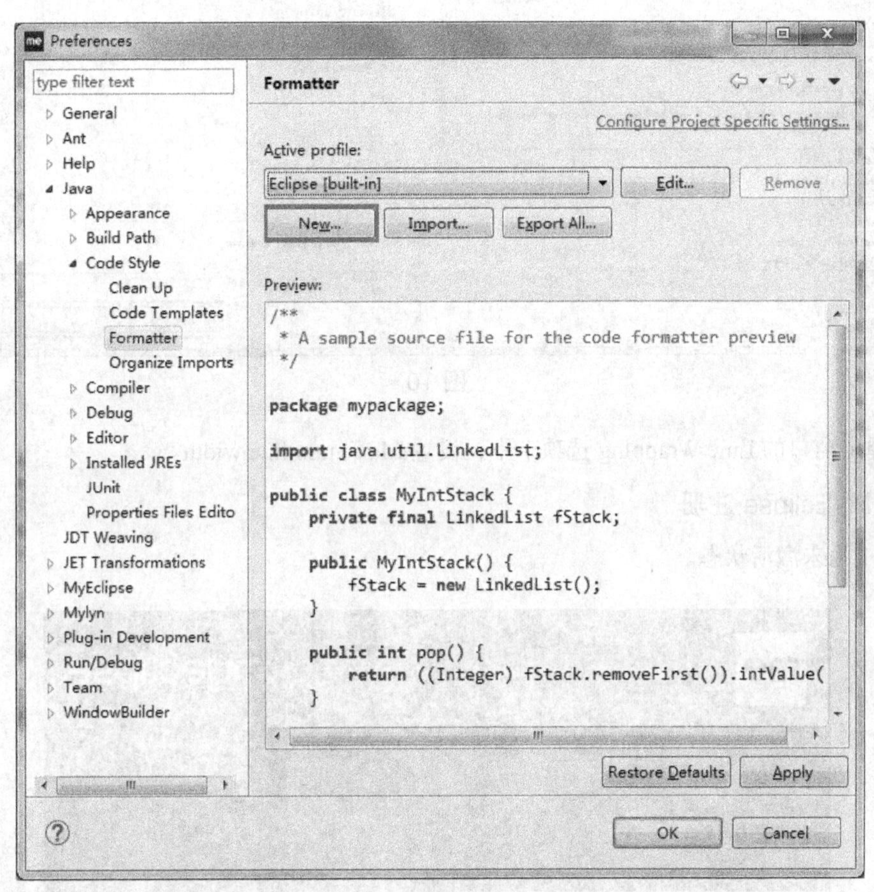

图 8

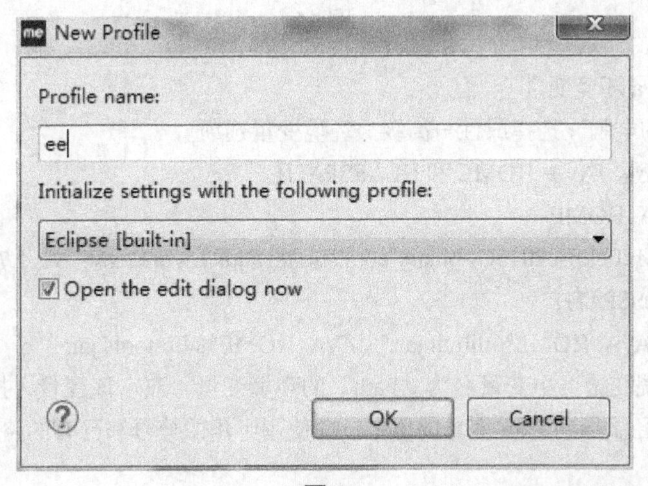

图 9

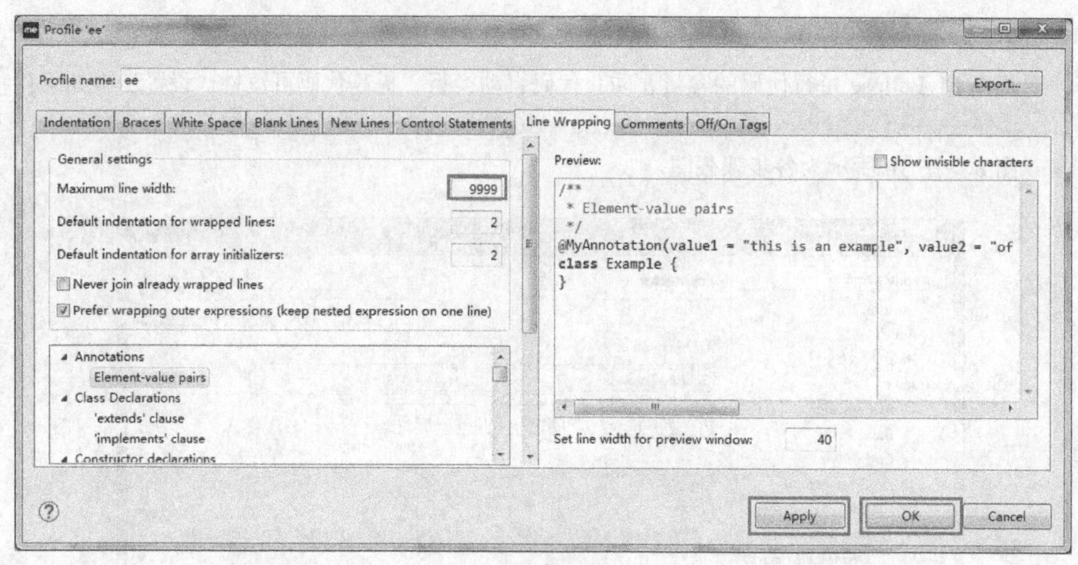

图 10

在弹出窗口的 Line Wrapping 选项卡中，设置 Maximum line width。

4. MyEclipse 注册

图 11 为未激活状态。

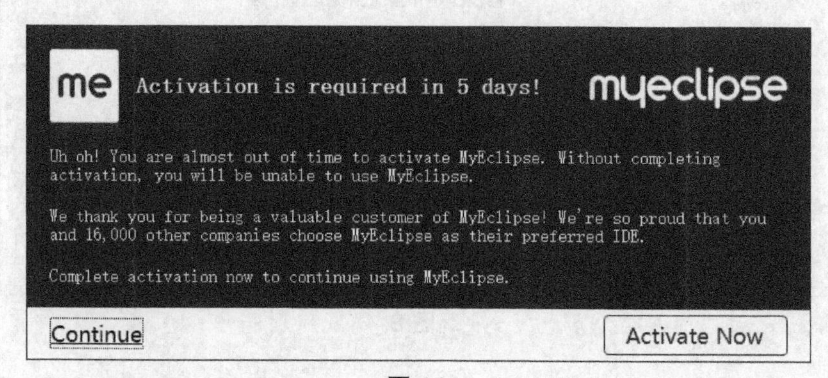

图 11

（1）设置 Java 环境变量

① 右击“我的电脑”选择属性→高级→环境变量选项。

② 新建系统变量 JAVA_HOME 和 CLASSPATH。

变量名：JAVA_HOME

变量值：C:\ MyEclipse2015CI\binary\com.sun.java.jdk7.win32.x86_64_1.7.0.u45

变量名：CLASSPATH

变量值：.;%JAVA_HOME%\lib\dt.jar;%JAVA_HOME%\lib\tools.jar;

③ 选择“系统变量”中变量名为“Path”的环境变量，双击该变量，把 JDK 安装路径中 bin 目录的绝对路径添加到 Path 变量的值中，并使用半角的分号和已有的路径进行分隔。

变量名：Path

变量值：%JAVA_HOME%\bin;%JAVA_HOME%\jre\bin;

（2）软件安装不要有中文目录。

（3）运行 crack.bat（见图 12）。

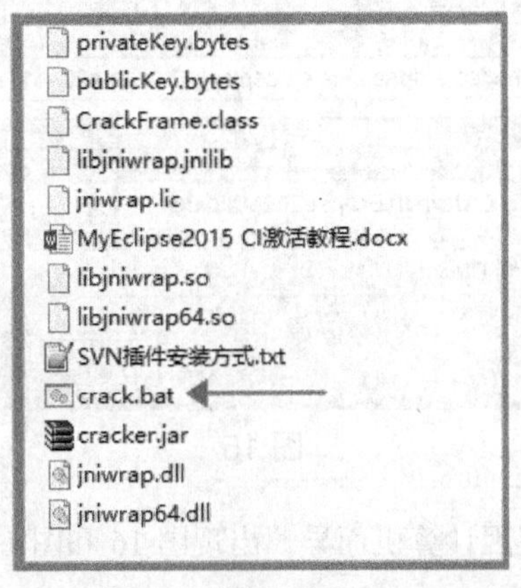

图 12

（4）在 UserCode 中输入本地计算机名（可右击“计算机”查看全名）。

（5）单击 SystemId，然后单击 Active（见图 13）。

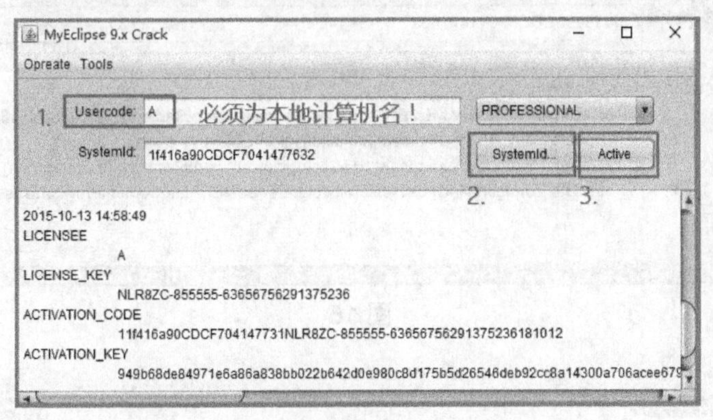

图 13

（6）单击菜单 Tools → 2.SaveProperties（见图 14）。

（7）单击菜单 Tools → 1.ReplaceJarFile（见图 14）。

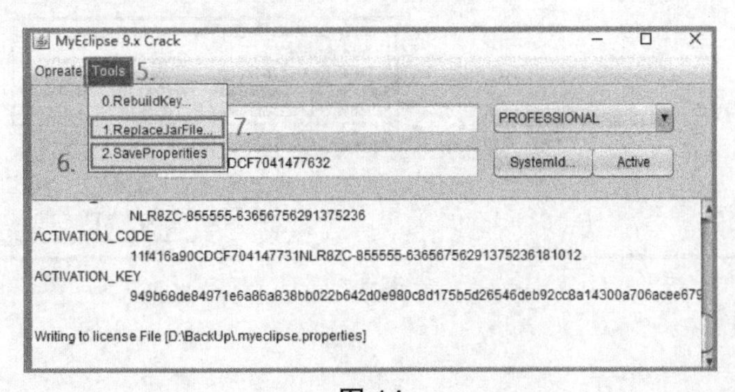

图 14

（8）选择 MyEclipse 2015 安装目录，然后单击打开（见图 15）。

Please Choose MyEclipse --> Plugin Folder
查找(I):
plugins
com.genuitec.eclipse.blue.ejbdeploy_13.0.0.me201411251616
文件夹名(N):
C:\Java\MyEclipse2015CI\plugins
8.
文件类型(T):
Dirctory
9.
打开(O)
取消

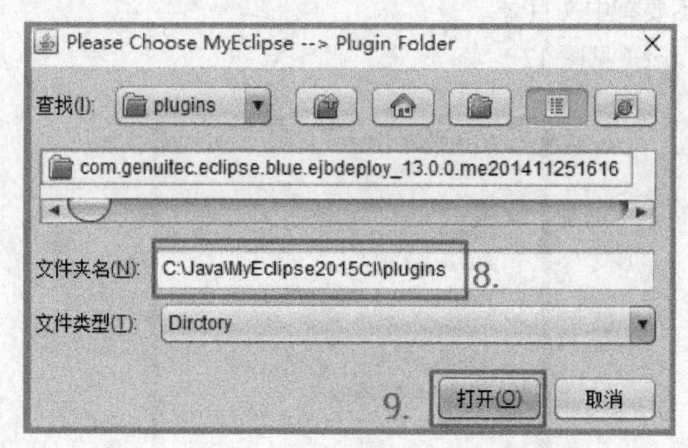

图 15

（9）稍等片刻，安装速度因计算机而异。出现图 16 和图 17 所示信息，表示完成激活。

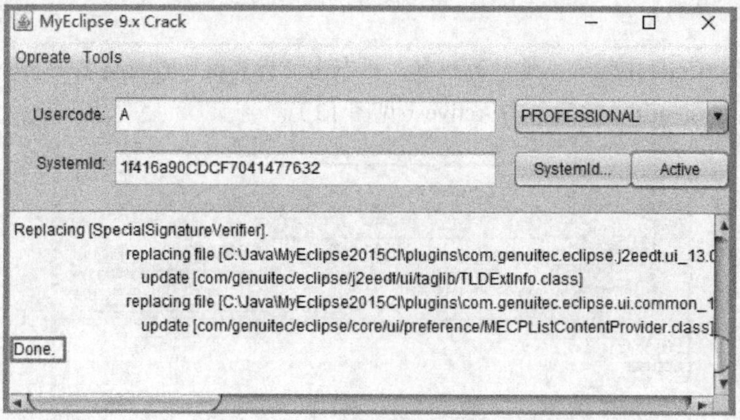

MyEclipse 9.x Crack
Opreate Tools
Usercode: A
PROFESSIONAL
SystemId: 1f416a90CDCF7041477632
SystemId...
Active
Replacing [SpecialSignatureVerifier].
replacing file [C:\Java\MyEclipse2015CI\plugins\com.genuitec.eclipse.j2eedt.ui_13.0
update [com/genuitec/eclipse/j2eedt/ui/taglib/TLDExtInfo.class]
replacing file [C:\Java\MyEclipse2015CI\plugins\com.genuitec.eclipse.ui.common_1
update [com/genuitec/eclipse/core/ui/preference/MECPListContentProvider.class]
Done.

图 16

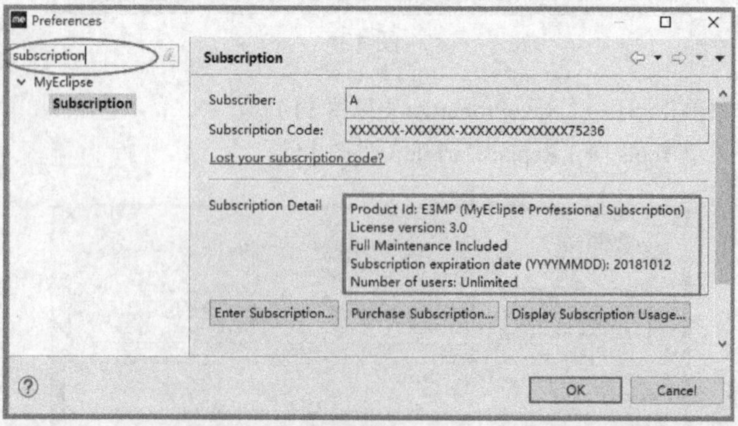

Preferences
subscription
MyEclipse
Subscription
Subscription
Subscriber: A
Subscription Code: XXXXXX-XXXXXX-XXXXXXXXXXXX75236
Lost your subscription code?
Subscription Detail
Product Id: E3MP (MyEclipse Professional Subscription)
License version: 3.0
Full Maintenance Included
Subscription expiration date (YYYYMMDD): 20181012
Number of users: Unlimited
Enter Subscription...
Purchase Subscription...
Display Subscription Usage...
OK
Cancel

图 17